PRINCIPLES OF MATHEMATICS

THIRD EDITION

SUSAN E. SWENEY ◫ **DOUGLAS B. COOK**

Owens Community College

KENDALL/HUNT PUBLISHING COMPANY
4050 Westmark Drive Dubuque, Iowa 52002

Cover image from Photos.com

Copyright © 1990, 1997, 2005 by Kendall/Hunt Publishing Company

ISBN 13: 978-0-7575-4006-6

All rights reserved. No part of this publication may be reproduced, stored in a retrieval system, or transmitted, in any form or by any means, electronic, mechanical, photocopying, recording, or otherwise, without the prior written permission of the copyright owner.

Printed in the United States of America
10   9   8   7   6

CONTENTS

Preface ix

CHAPTER 1   **WHOLE NUMBERS REVIEW** 1
                Introduction 1
              1-1   Basic Addition 2
              1-2   Addition of Larger Numbers 5
              1-3   Basic Subtraction 10
              1-4   Subtraction of Larger Numbers 13
              1-5   Basic Multiplication 19
              1-6   Multiplication of Larger Numbers 22
              1-7   Basic Division 28
              1-8   Division of Larger Numbers 31
                     Chapter 1 Review Tests 41

CHAPTER 2   **SYMBOLS AND DEFINITIONS** 45
              Introduction 45
              2-1   Whole and Decimal Numbers and Place Value 47
              2-2   Words Used for Mathematical Operations 50
              2-3   Factors, Prime and Composite Numbers 52
              2-4   Exponential Notation 53
              2-5   Order of Operations 55
              2-6   Commutative and Associative Properties 59
              2-7   Rational Numbers 62
              2-8   Procedure for Solving Application Problems 64
                     Chapter 2 Review Tests 69

CHAPTER 3   **PRIME NUMBERS** 73
              Introduction 73
              3-1   Listing All of the Factors of a Number 74
              3-2   Prime Versus Composite Numbers and Divisibility Tests 77
              3-3   Prime Factoring 83
              3-4   What Is a Lowest Common Multiple? 88
              3-5   Finding the Lowest Common Multiple 90
                     Chapter 3 Review Tests 101

              Cumulative Review #1–Chapters 1-3 105

CHAPTER 4 — FRACTIONS I: ADDITION AND SUBTRACTION 107

Introduction 107
- 4-1 What Is a Fraction? 108
- 4-2 Fraction Terms 109
- 4-3 Changing Improper Fractions into Mixed Numbers and Back Again 110
- 4-4 Reducing Fractions 114
- 4-5 Building Equivalent Fractions 119
- 4-6 Fraction Addition 123
- 4-7 Addition of Mixed Numbers 128
- 4-8 Fraction Subtraction 132
- 4-9 Mixed Number Subtraction and Borrowing 135

Chapter 4 Review Tests 143

CHAPTER 5 — FRACTIONS II: MULTIPLICATION AND DIVISION 147

Introduction 147
- 5-1 Multiplication of Fractions and Cancellation 148
- 5-2 Multiplication of Mixed Numbers 155
- 5-3 Division of Fractions 157
- 5-4 Division of Mixed Number Fractions 160

Chapter 5 Review Tests 163

CHAPTER 6 — DECIMALS I: ADDITION, SUBTRACTION, AND ROUNDING 167

Introduction 167
- 6-1 Place Value 168
- 6-2 Decimals to Fractions and Back Again 171
- 6-3 Rounding 176
- 6-4 Addition and Subtraction of Decimals 180

Chapter 6 Review Tests 185

Cumulative Review #2 – Chapters 1-6 189

CHAPTER 7 — DECIMALS II: MULTIPLICATION AND DIVISION 191

Introduction 191
- 7-1 Multiplication of Decimals 192
- 7-2 Shortcut for Multiplying by 10 or Powers of 10 195
- 7-3 Division of Decimals 198
- 7-4 Shortcut for Dividing by 10 or Powers of 10 202

Chapter 7 Review Tests 205

CHAPTER 8 — PERCENTS 209

Introduction 209
- 8-1 "Getting Out of Percent" 210
- 8-2 Changing Fractions and Decimals into Percents 217

Chapter 8 Review Tests 223

Arithmetic Review 227

Cumulative Review #3–Chapters 1-8 233

CHAPTER 9 INTRODUCTION TO THE METRIC SYSTEM 237
   Introduction 237
   9-1 Metric System and Calculator Method for Conversions
        in the Metric System 238
   9-2 Metric Prefix Conversion Line 241
   Chapter 9 Review Tests 243

CHAPTER 10 RATIOS, CONVERSIONS, AND PROPORTIONS 247
   Introduction 247
   10-1 Ratios 248
   10-2 Conversions 250
   10-3 Conversions with more than One Conversion Fraction 255
   10-4 Proportions 258
   10-5 More Proportions 261
   10-6 Applications of Proportions 264
   Chapter 10 Review Tests 269

CHAPTER 11 SYMBOLS OF PHARMACOLOGY 273
   Introduction 273
   11-1 Pharmacology Units 274
   Chapter 11 Review Tests 277

CHAPTER 12 SIGNED NUMBERS 279
   Introduction 279
   12-1 Multiplication of Signed Numbers 280
   12-2 Division of Signed Numbers 282
   12-3 More Information on Multiplication 284
   12-4 "Addition" of Like Signed Numbers 286
   12-5 "Addition" of Unlike (Different) Signed Numbers 290
   12-6 "Subtraction" of Signed Numbers 293
   12-7 "Addition" and "Subtraction" of Signed Numbers Using a Number Line .. 296
   12-8 Operations with Zero 306
   12-9 Order of Operations Reviewed 309
   12-10 Special Properties (Rules) Reviewed 314
   Chapter 12 Review Tests 317

Cumulative Review #4–Chapters 9-12 321

CHAPTER 13	**BASIC EQUATIONS IN ONE UNKNOWN** **323**
	Introduction 323
13-1	Solving Equations Such as $9x = 54$ 324
13-2	Solving Equations Such as $5x + 3 = 16$ 328
13-3	Solving Equations Such as $3x + 10 = 5x$ 332
13-4	Solving Equations Such as $5x - 3 = 2x + 12$ 335
13-5	Solving Equations Such as $5a - 3a + 3 = 6a - 4 + 9$ 339
13-6	Solving Equations Such as $\frac{4x}{3} = \frac{5}{6} - 2x$ 342
13-7	Solving Equations Such as $6(2x + 4) = 8(x - 5)$ 346
13-8	Review of Sections 13-1 through 13-7 350
13-9	Solving Equations Such as $m - \frac{m-2}{5} = \frac{2}{9}$ and $.6x - 6.32 = .04x + .4$ 352
	Chapter 13 Review Tests 355

CHAPTER 14	**LITERAL EQUATIONS** **359**
	Introduction 359
14-1	Solving Literal Equations 360
	Chapter 14 Review Tests 367

Cumulative Review #5–Chapters 9–14 371

CHAPTER 15	**POLYNOMIALS I: ADDITION, SUBTRACTION, AND GROUPING SYMBOLS** **375**
	Introduction 375
15-1	Monomials, Polynomials, and Like Terms .376
15-2	Combining Like Terms 379
15-3	Addition of Polynomials 381
15-4	Subtraction of Polynomials 383
15-5	Grouping Symbols 385
	Chapter 15 Review Tests 389

CHAPTER 16	**POLYNOMIALS II: MULTIPLICATION, DIVISION, AND EXPONENTS** **393**
	Introduction 393
16-1	Exponents 394
16-2	Multiplication of Polynomials 397
16-3	Monomial Times Polynomial 400
16-4	Polynomial Times Polynomial 402
16-5	Division of Polynomials 406
16-6	Monomial Divided by Monomial 408
16-7	Polynomial Divided by Monomial 412
16-8	Polynomial Divided by Polynomial 414
	Chapter 16 Review Tests 421

CHAPTER 17 GRAPHING **425**

          Introduction 425
    17-1   Introduction to Graphing 426
    17-2   Graphing Basic Equations—Lines and Parabolas 436
          Chapter 17 Review Tests 443

CHAPTER 18 SCIENTIFIC NOTATION **447**

          Introduction 447
    18-1   Exponents 448
    18-2   Multiplication Using Exponents 449
    18-3   Division Using Exponents 450
    18-4   Division (Continued) 452
    18-5   Significant Figures 454
    18-6   Rounding (A Number) 457
    18-7   Products, Quotients, and Rounding Based upon Significant Figures 459
    18-8   Scientific Notation (S.N.) 461
    18-9   Multiplication in Scientific Notation (S.N.) 463
    18-10 Division in Scientific Notation 465
    18-11 Multiplication and Division of Numbers Not in Scientific Notation 468
          Chapter 18 Review Tests 471

CHAPTER 19 SQUARE ROOTS **475**

          Introduction 475
    19-1   What is a Square Root? 476
    19-2   Key on Calculator 477
    19-3   Simplifying Square Roots 478
    19-4   Simplifying Higher Roots 483
          Chapter 19 Review Tests 488

    Algebraic Review 491

    Cumulative Review #6–Chapters 9-14, 18, 19 507

APPENDIX A ANSWERS TO EXERCISES AND REVIEW TESTS **511**

APPENDIX B SOLUTIONS TO ASTERISK PROBLEMS **591**

PREFACE

Principles of Mathematics has been designed for adult students who are enrolled in any numerical skills course. If you need a review of some basic arithmetic concepts, and if you need to "touch" on some basic algebra topics then this book *is* for you. All you will need to get through *Principles of Mathematics* is a few pencils, lots of paper and a willingness to do some work just enough to gain confidence in yourself and your ability to use math.

To help you in your efforts to do the work, a few special features have been included in this book.

FEATURE #1: In the back of *Principles of Math* you will find in Appendix A the answers to every exercise you will encounter in this book. As you work through a chapter completing the exercises, you can immediately see if you *are* understanding the concepts by checking your answers yourself.

FEATURE #2: In addition to all the answers to every exercise, there will also be some *solutions* to problems found in the exercises. Any exercise problem that has an asterisk (*) next to it will have its solution (how the answer was derived) in the answers portion of the book—Appendix B. This way if you don't get the right answer and can't find an example problem in the book like the exercise problem you're working on, look at the solutions to other problems in the exercise and see if that doesn't help you come up with the answer. Of course, if you still don't see the answer, *ask your instructor for help.* It is possible that your answer is the correct one, and the book is wrong (nobody is perfect, but we've tried to keep our mistakes down to a minimum). If the book *is* wrong, have your instructor make a note of it so we can correct our mistake in the future.

FEATURE #3: At the end of every chapter you will find *Chapter Review Tests.* These tests are as much like the topic quizzes that you will be taking over *your* required chapters as possible. This will better prepare you for the quiz when you are ready to take it. (You ask your instructor for topic quizzes.)

FEATURE #4: You will find *Cumulative Review Tests* throughout this book. These tests are designed to help you review and retain the math skills you have learned. It is *vitally important* to complete these review tests as you move through this book. As with exercises, the answers will be in Appendix A.

FEATURE #5: After the *Percent* chapter (Chapter 8), you will find some *Arithmetic Review* pages. These are designed to give you a *very quick* review of the arithmetic chapters. These pages are there to help you review for the Arithmetic Proficiency Test that you will take from your instructor after you complete the *Percent* chapter.

Also, at the end of the *Square Roots* chapter, you will find an *Algebraic Review* which gives you a quick review of the algebra topics, to help you prepare for the Algebraic Proficiency Test which will also be given to you by your instructor.

FEATURE #6: This feature is not a feature of your book; this feature is your *instructor.* If at any time you find yourself lost or confused about the math you are studying, *ask your instructor* for help! That is precisely what the instructor is there for, so put her/him to work!

As you study through this book remember one thing: YOU ARE ONE OF THE MOST IMPORTANT PEOPLE HERE! So, *thank you* for choosing us to help you achieve your goals, and here's to your success in mathematics and beyond . . .

Susan E. Sweney
Douglas B. Cook

CHAPTER 1

WHOLE NUMBERS REVIEW

INTRODUCTION

This chapter is designed to give a quick review of the whole number operations. As you work through this chapter keep in mind that you are not only working the problems to get the correct answer, but you are also trying to increase your speed at getting the correct answer. You must know the *basic* whole number facts almost without consciously thinking about them. These basic operations are the foundation for all the math to be learned in the future. Can you picture what would happen to the Empire State Building if it did not have a solid foundation? Disaster! So remember: *Your* solid foundation for future success in mathematics is accuracy and speed with the whole number facts!

SECTION 1-1 BASIC ADDITION

Basic Addition Facts #1

Time Yourself

1 +2	3 +5	9 +1	2 +1	3 +6	5 +2	8 +0	9 +3	7 +2	3 +9
3 +8	7 +4	5 +6	8 +7	0 +3	1 +7	2 +9	3 +8	6 +9	8 +6
7 +7	3 +8	9 +5	6 +6	8 +9	0 +6	4 +7	6 +5	8 +8	7 +3
1 +1	8 +2	7 +3	4 +9	4 +8	6 +7	3 +3	7 +8	9 +9	9 +6
5 +8	6 +6	3 +5	4 +0	9 +8	7 +6	8 +7	9 +5	7 +4	5 +5
8 +7	9 +8	6 +6	7 +9	4 +8	3 +9	5 +5	8 +5	6 +7	2 +3
3 +8	3 +1	4 +2	7 +7	8 +4	0 +9	7 +0	6 +6	4 +5	8 +0
9 +3	5 +8	7 +7	6 +2	9 +9	8 +8	5 +4	7 +6	8 +6	9 +7
0 +0	2 +9	4 +4	8 +7	9 +8	7 +9	6 +1	3 +8	9 +4	7 +1
6 +5	4 +8	7 +9	3 +5	2 +1	0 +8	8 +8	6 +7	6 +6	7 +9

Check your time and answers! If you took longer than five minutes and/or missed more than ten problems, you will want to study the problems you missed and take **Basic Addition Facts #2**. If you missed ten or fewer problems and completed **Basic Addition Facts #1** within five minutes, study the facts you missed then continue on to **Addition of Large Numbers #1**.

Basic Addition Facts #2

Time Yourself

3 +9	7 +2	8 +0	9 +3	5 +2	3 +6	2 +1	1 +9	3 +5	1 +2
8 +6	6 +9	3 +8	2 +9	1 +7	0 +3	8 +7	5 +6	7 +4	3 +8
7 +3	8 +8	6 +5	4 +7	0 +6	8 +9	6 +6	9 +5	3 +8	7 +7
9 +6	9 +9	7 +8	3 +3	6 +7	4 +8	4 +9	7 +3	8 +2	1 +1
5 +5	7 +4	9 +5	8 +7	7 +6	9 +8	4 +0	3 +5	6 +6	5 +8
2 +3	6 +7	8 +5	5 +5	3 +9	4 +8	7 +9	6 +6	9 +8	8 +7
8 +0	4 +5	6 +6	7 +0	0 +9	8 +4	7 +7	4 +2	3 +1	3 +8
9 +7	8 +6	7 +6	5 +4	8 +8	9 +9	6 +2	7 +7	5 +8	9 +3
7 +1	9 +4	3 +8	6 +1	7 +9	9 +8	8 +7	4 +4	2 +9	0 +0
6 +5	7 +9	6 +6	6 +7	8 +8	0 +8	2 +1	3 +5	7 +9	4 +8

Check your *time* and *answers.* If you took longer than five minutes and/or missed more than ten problems, study the problems you missed and take **Basic Addition Facts #3**. If you missed ten or fewer problems and completed **Basic Addition Facts #2** within five minutes, study the facts you missed then continue on to **Addition of Larger Numbers #1**.

Basic Addition Facts #3

Time Yourself

7	7	3	6	7	0	8	3	2	4
+1	+9	+8	+7	+9	+8	+7	+5	+9	+8

6	9	6	6	8	9	2	4	7	0
+5	+4	+6	+1	+8	+8	+1	+4	+9	+0

8	8	6	5	0	9	7	4	5	3
+0	+6	+6	+4	+9	+9	+7	+2	+8	+8

9	4	7	7	8	8	6	4	3	9
+7	+5	+6	+0	+8	+4	+2	+2	+1	+3

5	6	9	3	1	8	3	9	7	3
+5	+7	+5	+5	+2	+7	+9	+8	+9	+5

9	5	2	7	8	5	7	4	4	6
+8	+8	+3	+4	+5	+5	+6	+8	+0	+6

9	8	7	9	6	3	0	4	6	7
+2	+7	+3	+9	+5	+3	+6	+8	+6	+3

3	1	9	8	7	4	6	8	4	9
+8	+1	+6	+8	+8	+7	+7	+9	+9	+5

8	7	8	6	3	2	1	0	8	5
+2	+7	+6	+9	+8	+9	+7	+3	+7	+6

7	3	3	7	8	9	5	3	2	1
+4	+8	+9	+2	+0	+3	+2	+6	+1	+9

Check your *time* and your *answers*. If you took longer than five minutes and/or missed more than ten problems, it's *important* that you do *not* continue on until you have spoken to your instructor. If you missed ten or fewer problems and completed **Basic Addition #3** in five minutes or less, study the facts you missed then you may continue on to **Addition of Larger Numbers #1**.

SECTION 1-2 ADDITION OF LARGER NUMBERS

1.   2380
        956
    +   83

Add first column

$$\begin{array}{r}\downarrow\\ 2380\\ 956\\ +83\\ \hline 9\end{array}$$

Add second column   Indicate "carry-over"

$$\begin{array}{r}2\downarrow\\ 2380\\ 956\\ +83\\ \hline 19\end{array}$$

Add third column   Indicate next "carry-over"

$$\begin{array}{r}\downarrow\\ \mathbf{1\,2}\\ 2380\\ 956\\ +83\\ \hline 419\end{array}$$

Add fourth column

$$\begin{array}{r}\downarrow\\ \mathbf{1}\\ 2380\\ 956\\ +83\\ \hline 3419\end{array}\;=\;\mathbf{Answer}$$

2.   99876
      39752
      83648
    + 27965

Add first column   Indicate "carry-over"

$$\begin{array}{r}2\downarrow\\ 99876\\ 39752\\ 83648\\ +\,27965\\ \hline 1\end{array}$$

Add next column   Indicate next "carry-over"

$$\begin{array}{r}\downarrow\\ \mathbf{2\,2}\\ 99876\\ 39752\\ 83648\\ +\,27965\\ \hline 41\end{array}$$

	↓	
	32	Indicate next "carry-over"
Add next column	99876	
	39752	
	83648	
	+ 27965	
	241	

	↓	
	33	Indicate next "carry-over"
Add next column	99876	
	39752	
	83648	
	+ 27965	
	1241	

	↓	
	3	Indicate next "carry-over"
Add next column	99876	
	39752	
	83648	
	+ 27965	
	251241 = **Answer**	

Addition of Larger Numbers #1

8	9	7	8	4	3	7	5	1	9
3	2	6	9	7	9	6	5	9	9
+4	+7	+7	+7	+8	+2	+9	+3	+6	+9

18	22	19	27	35	43	87	93	65	76
+7	+8	+2	+6	+7	+9	+8	+7	+6	+9

22	34	95	34	72	85	67	93	52	37
+12	+18	+21	+34	+36	+91	+82	+48	+49	+16

237	159	842	761	637	830	427	988	749	813
+65	+39	+75	+99	+95	+75	+66	+78	+77	+29

71690	62358	4796	33521	8431	9867	2000	7401	8003	1999
8210	9756	2950	6879	1100	2001	1749	1345	4024	896
+379	+4387	+1739	+693	+900	+1899	+386	+592	+4097	+753

8931	2168	2000		6395	453	389		90010	38092
4526	347	109		8790	1598	4000		2359	48103
387	982	380		291	231	553		867	61300
6435	36	6475		13	8460	68		1542	21509
+93	+101	+18	+6	+93	+17			+180	+41000

16	23	89	99	3	17	10	15	21	33
18	47	68	82	4	42	20	25	22	32
23	85	43	73	5	38	30	35	23	31
19	16	56	65	6	22	40	45	24	30
7	10	48	18	7	66	50	55	25	29
6	6	7	19	8	54	60	65	26	28
+5	+3	+2	+20	+9	+98	+70	+75	+27	+27

Check your answers. If you missed more than five problems, find your mistakes, correct them, and do **Addition of Larger Numbers #2**; otherwise, correct your mistakes and continue on to **Basic Subtraction #1**.

Addition of Larger Numbers #2

```
   3      9      5      1      8      6      3      2      9      6
   7      2      7      9      4      8      8      4      9      7
 + 8    + 6    + 3    + 9    + 7    + 9    + 7    + 6    + 8    + 8


  38     19     23     74     83     65     34     57     42     77
 + 2    + 7    + 8    + 9    + 6    + 8    + 9    + 9    + 8    + 6


  29     86     39     48     79     35     23     97     66     51
 +35    +74    +27    +83    +68    +47    +88    +39    +45    +79


 100    687    221    605    911    763    359    873    747    456
 + 99   + 93   + 89   + 37   + 89   + 48   + 55   + 29   + 90   + 87


 2400   8792   1000   7010   3999  47835   8709   3200   5236    711
  301    632    985    838    499   2359    709    999   7454   1528
 + 979  + 740  +  63  +  96  + 159  +7431  + 811  + 666  +3832  + 376


 2986   1586   7600   4441   9090   8912  39051  10000
 1000    280   1800    359    100   3943  48502   8001
  387   4368    366   6601    867   6951   1698   3581
  452     42    475    870    329   8000   7500   6900
 +  6   + 11   + 892  +  38  +1061  + 600  + 301  +3500


  12     32      3     89     63     78      8     10     83     22
  10     27      7     72     27     79      9     19     27     34
   9     16      9     36     82     81     17     20     90     46
   8      9      8     54     19     63     16     29     99     58
   7      7      3     29     23     42     20     30     16     60
   6      5      2     18     86     19     37     39      2     72
 + 5    + 2    + 1    +17    +99    +21    +40    +40    + 1    +74
```

Check your answers. If you missed more than five problems, find your mistakes, and correct them, and do **Addition of Larger Numbers #3**; otherwise, correct your mistakes and go on to **Basic Subtraction #1**.

Addition of Larger Numbers #3

```
   6       8       9       6       3       7       5       8       3       4
   2       3       8       7       6       7       4       8       3       4
  +9      +7      +4      +5      +8      +3      +9      +6      +9      +7


  38      42      87      63      39      41      22      89      68      91
  +8      +6      +9      +5      +7      +3      +8      +7      +5      +9


  69      82      55      78      86      91      22      36      98      10
 +24     +39     +63     +89     +88     +93     +57     +44     +88     +32


 385     459     698     297     800     684     100     258     566     997
 +26    + 87   + 100   + 309   + 301   + 793   + 900   + 372   + 733   + 856


 600    7891    4009    3999   86010     679   10000   98976    7001     279
 704    2349    1008     492    3795     832    4932    5432     899     348
+809   +8799   + 201   +  34   + 430   + 476   +  48   + 345   + 642   + 156


3892    9881    7388    4532    5000    6111   10008           80000
6010     684     201     899     300     833   11526           17344
 822    7290     398     522      20    7008   42378            6598
 156    6310     431     600       9     697    5973             432
+ 17   + 542   + 846   + 418   +  80   + 214   +   29         +    61


   8      11      21      29      48      30      20      45      79      63
   7      26      21      74      33      42      13      60      89      27
   3      73      32      56      15      89      15      70      99      98
   4      85      47      83      77      93      28      80       9      45
   9       4      86      21      49      67      69      96      19      18
   6       7       9      18      80      43      73      21      29      23
 + 1     + 8     + 3    + 17    + 32    + 7    + 28    + 43    + 39    + 73
```

 Check your answers. If you have missed more than five problems, it is important that you speak with your instructor *before* continuing on. If you missed five or fewer problems, correct your mistakes then you may continue on to **Basic Subtraction #1**.

SECTION 1-3 BASIC SUBTRACTION

Basic Subtraction Facts #1

Time Yourself

1	4	2	5	6	8	7	9	3	4
−1	−3	−0	−4	−5	−1	−5	−8	−1	−4

3	8	9	7	8	7	8	6	2	4
−2	−8	−0	−6	−3	−4	−2	−4	−2	−1

6	8	9	7	4	9	7	5	4	0
−0	−7	−3	−6	−2	−9	−1	−3	−2	−0

5	9	14	9	18	16	17	9	14	16
−0	−5	−6	−7	−9	−8	−9	−7	−9	−7

13	12	15	17	11	11	12	16	14	17
−4	−8	−6	−8	−3	−2	−5	−9	−8	−8

15	13	11	11	14	12	16	12	12	11
−7	−6	−2	−3	−9	−3	−7	−6	−8	−3

10	15	10	10	14	12	15	18	14	11
−1	−7	−2	−3	−5	−4	−9	−9	−5	−3

12	13	11	9	15	8	6	13	14	10
−3	−6	−4	−0	−6	−2	−0	−7	−5	−2

13	10	11	12	10	12	16	12	12	15
−9	−1	−3	−3	−3	−4	−7	−7	−3	−6

11	8	12	14	14	17	12	11	13	13
−3	−0	−9	−8	−5	−9	−5	−5	−6	−7

Check your *time* and your *answers*. If you missed more than ten problems and/or took longer than five minutes, you will want to study the problems you missed and take **Basic Subtraction #2**. If you missed fewer than ten problems and took five minutes or less to complete **Basic Subtraction #1**, correct your mistakes then continue on to **Subtraction of Larger Numbers #1**.

Basic Subtraction Facts #2

Time Yourself

0 − 0	8 − 1	7 − 3	1 − 1	2 − 1	3 − 0	6 − 4	8 − 5	9 − 1	7 − 5
3 − 1	3 − 3	5 − 2	6 − 6	7 − 2	8 − 6	9 − 4	4 − 4	0 − 0	2 − 1
9 − 6	8 − 7	6 − 1	5 − 3	3 − 1	9 − 2	8 − 3	6 − 5	6 − 3	9 − 7
8 − 4	9 − 3	7 − 4	6 − 2	5 − 4	4 − 2	3 − 2	9 − 5	8 − 7	7 − 7
6 − 0	7 − 7	8 − 2	9 − 9	5 − 3	4 − 1	9 − 0	8 − 8	7 − 4	9 − 3
13 − 4	12 − 3	14 − 6	16 − 9	12 − 5	11 − 2	11 − 5	10 − 1	13 − 4	10 − 3
10 − 5	11 − 2	16 − 9	16 − 7	16 − 8	15 − 6	11 − 4	11 − 2	11 − 3	15 − 7
13 − 8	12 − 5	13 − 7	12 − 3	11 − 3	10 − 1	10 − 3	14 − 6	17 − 8	14 − 8
12 − 3	16 − 7	13 − 9	12 − 8	9 − 0	11 − 5	13 − 9	16 − 7	10 − 3	16 − 9
12 − 4	13 − 9	11 − 6	11 − 2	10 − 3	13 − 6	15 − 7	14 − 8	13 − 6	15 − 6

Check your time and answers. If you missed more than ten problems, and/or took longer than five minutes to complete the problems, study your mistakes and continue on to **Basic Subtraction #3**. Otherwise you may continue on to **Subtraction of Larger Numbers #1**.

Basic Subtraction Facts #3

Time Yourself

4 −0	5 −1	6 −5	3 −2	8 −1	7 −0	9 −8	3 −1	5 −5	2 −2
8 −6	6 −4	5 −3	2 −1	0 −0	9 −7	7 −2	4 −3	3 −3	1 −1
9 −3	6 −1	8 −7	5 −4	7 −5	4 −2	9 −6	8 −5	3 −0	2 −0
4 −1	6 −3	9 −0	8 −4	7 −3	0 −0	3 −2	7 −6	7 −4	8 −2
9 −2	8 −3	7 −6	6 −5	0 −0	9 −7	1 −0	8 −1	7 −0	9 −1
10 −2	13 −7	11 −3	18 −9	10 −1	9 −0	14 −6	11 −2	12 −4	10 −3
18 −9	16 −7	14 −8	12 −5	11 −2	12 −3	14 −7	12 −4	13 −6	16 −9
11 −4	10 −3	15 −6	13 −7	12 −3	11 −4	17 −8	9 −3	16 −8	11 −4
11 −5	12 −4	10 −2	14 −8	13 −8	15 −6	13 −6	14 −5	12 −6	11 −2
18 −9	12 −4	16 −9	11 −4	13 −8	13 −5	12 −9	14 −9	12 −4	16 −9

Check your time and answers. If you took longer than five minutes, and/or missed more than ten problems, then see your instructor as soon as possible. Do not continue on. If you took five minutes or less to complete **Basic Subtraction #3**, and you missed less than ten problems, correct your mistakes then continue on to **Subtraction of Larger Numbers #1**.

SECTION 1-4 SUBTRACTION OF LARGER NUMBERS

1. 3962
    − 798

Try to subtract first column

$\downarrow$
3962
− 798

2 − 8 = ?
Cannot do

If you cannot subtract first column, borrow—then subtract

$\downarrow$
5_1
39͟62
− 798
‾
   4

12 − 8 = 4

Try to subtract next column

$\downarrow$
5
396͟2
− 798
‾
   4

5 − 9 = ?
Cannot do

Borrow, then subtract

$\downarrow$
^{1}
8 5
3͟9͟62
− 798
‾
  64

15 − 9 = 6

Try to subtract next column

$\downarrow$
8
3͟962
− 798
‾
 164

8 − 7 = 1

Subtract next column

$\downarrow$
3962
− 798
‾
3164 = **Answer**

3 − 0 = 3

CHECK: 3164
          + 798
          ‾‾‾‾‾
             3962

2. 4000
   − 392

	Try to subtract first column	↓ 4000 − 392	0 − 2 = ? **Cannot do**
	Try to borrow	⌒ 4000 − 392	*Cannot* borrow from 0. Try next column.
	Try to borrow	⌒ 4000 − 392	Still cannot borrow from 0. Try next column.
	Try to borrow	3 ⌒ 4̸000 − 392	Borrow 1 from 4.
		3₁ 4̸000 − 392	Add 10 to column *next* to 4.
	Borrow	3₁9₁ 4̸0̸00 − 392	Borrow from 10.
		3₁9₁ 4̸0̸00 − 392	Add 10 to next column.
	Borrow	3₁9₁9 4̸0̸0̸0 − 392	Borrow from 10.
	Borrow	3̸9̸9̸₁ 4000 − 392	Add 10 to next column.
	Subtract	3₁9₁9₁ 4̸0̸0̸0 − 392 3608	= **Answer**

CHECK: 3608
           + 392
           —————
             4000

3. Try to subtract first column

$$\begin{array}{r}\downarrow\\102031\\-9136\end{array}\Big]$$

1 − 6 = ?
Cannot do

Borrow

$$\begin{array}{r}{}^{2_1}\\10203\!\!\!/1\\-9136\end{array}$$

Borrow and add 10 to column.

Subtract first column

$$\begin{array}{r}\downarrow\\{}^{2_1}\\10203\!\!\!/1\\-9136\\\hline5\leftarrow\end{array}$$

11 − 6 = 5

Try to subtract next column

$$\begin{array}{r}\downarrow\\2\\10203\!\!\!/1\\-9136\\\hline5\end{array}$$

2 − 3 = ?
Cannot do

Borrow

$$\begin{array}{r}\downarrow\\1_1\,2\\1020\!\!\!/3\!\!\!/1\\-9136\\\hline5\end{array}$$

Borrow and add 10 to column.
NOTICE: Cannot borrow from 0, so go to next column.

Try to subtract next column

$$\begin{array}{r}\downarrow\\1\\102\!\!\!/031\\-9136\\\hline895\end{array}$$

1 − 9 = ?
Cannot do

Borrow

$$\begin{array}{r}\downarrow\\{}^{1}\\0_1\,9\,1\\1\!\!\!/0\!\!\!/2031\\-9136\\\hline895\end{array}$$

Borrow and add 10 to column.

Borrow
Subtract

$$\begin{array}{r}\downarrow\downarrow\\0\;9\,11\\1\!\!\!/0\!\!\!/2031\\-9136\\\hline92895\end{array}$$ = **Answer**

Borrow from 10 and add to next column.
11 − 9 = 2. Finish subtracting.

CHECK:
$$\begin{array}{r}92895\\+9136\\\hline 102031\end{array}$$

REMEMBER: You *cannot* borrow from zero (0). You must go to the next column which *has* a number other than zero and borrow from that column.

Subtraction of Larger Numbers #1

25	37	86	97	110	87	53	49	75	19
−15	−22	−43	−64	−10	−26	−22	−36	−63	−12

36	27	47	53	88	63	28	70	32	18
−9	−8	−9	−7	−9	−5	−9	−9	−9	−9

26	31	40	63	50	87	60	72	31	54
−8	−9	−6	−4	−9	−8	−3	−4	−6	−9

70	80	60	100	350	500	251	300	452	600
−19	−28	−15	−29	−48	−67	−88	−76	−98	−94

806	367	747	896	1029	987	398	677	459	531
−102	−152	−325	−781	−1023	−265	−365	−526	−323	−200

382	683	726	854	800	956	892	541	325	276
−97	−95	−57	−75	−73	−72	−93	−51	−34	−92

151	223	815	1001	2009	8000	507	691	200	330
−69	−47	−71	−98	−61	−28	−42	−23	−53	−61

853	428	897	731	296	438	4000	5020	7300	3802
−725	−336	−698	−456	−137	−359	−296	−618	−369	−891

6070	1321	789	1789	5222	4630	2000	4195	3000	5009
−594	−163	−359	−769	−833	−647	−896	−296	−210	−673

10000	16500	8120	20000	76522	70000	28500	36590
−8500	−7652	−6256	−19480	−8461	−69999	−16701	−28463

Check your answers. If you missed more than five problems, study the problems you missed, make your corrections and proceed to **Subtraction of Larger Numbers #2.** If you missed five or fewer problems, correct your mistakes and continue on to **Basic Multiplication #1.**

Subtraction of Larger Numbers #2

28	36	42	63	75	61	54	23	16	25
− 9	− 7	− 5	− 8	− 9	− 6	− 7	− 7	− 7	− 6

82	47	35	81	77	34	27	63	38	76
− 9	− 8	− 7	− 8	− 9	− 8	− 8	− 4	− 9	− 7

83	62	98	75	67	81	60	36	49	54
− 31	− 21	− 63	− 32	− 46	− 20	− 50	− 25	− 38	− 13

65	83	45	88	77	32	60	70	50	61
− 29	− 67	− 26	− 69	− 48	− 19	− 21	− 38	− 43	− 32

800	670	473	229	846	731	693	821	756	801
− 47	− 79	− 78	− 39	− 58	− 38	− 97	− 64	− 98	− 61

726	700	900	856	732	890	675	893	437	221
− 49	− 61	− 99	− 29	− 43	− 91	− 75	− 84	− 29	− 38

108	632	958	706	555	283	479	892	999	325
− 107	− 420	− 637	− 402	− 442	− 160	− 475	− 372	− 637	− 104

392	675	387	541	238	800	732	950	751	837
− 187	− 598	− 293	− 451	− 149	− 351	− 600	− 496	− 692	− 138

8000	9001	832	698	852	9000	275	693	3092	426
− 436	− 721	− 741	− 698	− 763	− 852	− 174	− 594	− 2832	− 419

10000	2459	7500	20000	180001	8903	7508	13695
− 9863	− 2368	− 6960	− 6325	− 29002	− 604	− 4936	− 2985

Check your answers. If you missed more than five problems, study those problems and continue on to **Subtraction of Larger Numbers #3**. Otherwise, correct your mistakes and continue on to **Basic Multiplication #1**.

Subtraction of Larger Numbers #3

32 − 8	47 − 9	63 − 5	87 − 8	91 − 8	65 − 7	32 − 7	43 − 6	71 − 5	24 − 7
50 − 3	68 − 9	24 − 6	85 − 6	27 − 7	32 − 6	75 − 8	69 − 9	85 − 9	30 − 4
82 − 61	38 − 15	59 − 44	20 − 20	35 − 15	63 − 22	88 − 67	95 − 62	87 − 26	44 − 34
55 − 26	97 − 89	61 − 59	48 − 39	22 − 19	63 − 48	95 − 38	67 − 28	21 − 16	47 − 29
152 − 63	875 − 49	750 − 61	980 − 65	735 − 46	900 − 72	101 − 36	555 − 56	738 − 46	901 − 35
860 − 73	900 − 87	392 − 95	879 − 79	246 − 57	800 − 24	626 − 47	832 − 33	720 − 19	357 − 48
152 − 121	687 − 532	747 − 126	963 − 751	859 − 648	267 − 143	486 − 132	327 − 123	961 − 850	380 − 260
738 − 543	279 − 183	700 − 658	395 − 296	750 − 123	999 − 890	396 − 296	720 − 635	523 − 485	233 − 143
8000 − 452	9001 − 638	5401 − 601	7503 − 934	2000 − 1000	1856 − 955	7270 − 281	3000 − 295	4700 − 607	5211 − 322
25678 − 7692	38004 − 18905	75021 − 6592	100000 − 38920	70000 − 3960	45021 − 6000	23941 − 19852	6000 − 6000		

Check your answers. If you missed more than five problems, see your instructor as soon as possible for further assistance. *Do not continue on.* If you missed five or fewer problems, correct your mistakes and continue on to **Basic Multiplication #1**.

SECTION 1-5 BASIC MULTIPLICATION

Basic Multiplication #1

Time Yourself

5	7	9	5	5	6	6	9	8	8
×2	×2	×2	×3	×4	×3	×4	×4	×2	×4

3	2	7	6	4	5	5	9	3	8
×5	×5	×3	×5	×8	×9	×8	×5	×7	×5

3	2	6	9	9	5	5	7	8	7
×4	×8	×3	×4	×3	×6	×4	×5	×3	×4

3	4	9	3	5	7	7	4	9	4
×7	×6	×5	×5	×8	×3	×2	×8	×3	×9

5	6	7	8	3	4	9	6	7	6
×6	×6	×7	×8	×9	×7	×5	×7	×5	×5

9	9	9	9	8	8	8	8	8	5
×9	×7	×5	×3	×2	×6	×4	×8	×5	×9

9	8	6	6	7	4	0	7	6	9
×6	×9	×5	×7	×8	×9	×5	×5	×9	×0

5	8	8	7	4	8	7	9	8	6
×7	×4	×8	×3	×5	×6	×9	×6	×7	×8

9	8	9	5	2	3	0	9	0	7
×7	×5	×9	×5	×1	×0	×0	×6	×7	×2

3	8	7	8	9	6	8	9	0	9
×9	×7	×7	×0	×6	×7	×9	×9	×3	×8

Check your *time* and *answers*. If you missed more than ten problems, and/or took longer than five minutes to complete these problems, then study the missed problems and take **Basic Multiplication #2**. Otherwise, study the facts you missed and continue on to **Multiplication of Larger Numbers #1**.

Basic Multiplication #2

Time Yourself

7 × 2	5 × 3	6 × 4	9 × 4	7 × 3	5 × 2	5 × 3	6 × 3	1 × 2	3 × 1
3 × 5	2 × 7	7 × 3	4 × 6	8 × 4	9 × 5	5 × 8	5 × 9	3 × 2	2 × 2
3 × 4	2 × 8	6 × 3	9 × 4	9 × 2	5 × 4	6 × 5	5 × 7	6 × 4	9 × 3
0 × 0	3 × 7	9 × 5	4 × 6	3 × 5	8 × 5	7 × 3	2 × 7	4 × 8	0 × 9
5 × 6	6 × 6	7 × 7	8 × 8	3 × 9	4 × 7	5 × 9	9 × 5	6 × 7	8 × 0
9 × 9	7 × 9	8 × 2	9 × 5	4 × 8	9 × 3	6 × 8	8 × 8	7 × 8	3 × 3
9 × 6	9 × 8	6 × 5	7 × 6	8 × 7	4 × 9	6 × 9	5 × 7	8 × 6	4 × 4
3 × 8	9 × 0	6 × 8	5 × 7	4 × 8	7 × 3	5 × 4	5 × 5	8 × 6	7 × 9
9 × 6	8 × 0	6 × 6	3 × 9	8 × 7	6 × 9	5 × 4	6 × 8	0 × 3	4 × 1
7 × 7	8 × 3	9 × 9	6 × 7	7 × 6	5 × 3	1 × 5	6 × 8	9 × 0	7 × 9

Check your time and answers. If you missed more than ten problems, and/or took longer than five minutes to complete the problems, then study the multiplication facts that you missed and proceed to **Basic Multiplication #3**. Otherwise, study the facts you missed and continue on to **Multiplication of Larger Numbers #1**.

Basic Multiplication #3

Time Yourself

3	4	9	3	2	3	4	2	3	4
×7	×6	×5	×5	×1	×0	×2	×2	×1	×2

9	5	5	7	5	6	3	9	6	3
×2	×2	×3	×2	×4	×3	×4	×4	×4	×5

7	7	3	7	6	4	6	5	4	9
×2	×4	×4	×3	×4	×8	×5	×5	×4	×0

3	2	6	9	9	5	6	5	0	7
×4	×8	×3	×4	×2	×6	×6	×4	×8	×5

0	5	7	8	3	9	4	9	4	6
×0	×6	×7	×8	×9	×9	×8	×4	×7	×7

9	8	9	7	3	4	6	5	7	6
×8	×8	×0	×7	×9	×8	×0	×4	×7	×7

9	8	6	6	9	7	4	7	6	6
×6	×9	×5	×7	×6	×8	×9	×5	×0	×9

5	8	8	7	4	8	6	7	9	8
×7	×4	×8	×3	×5	×6	×6	×9	×6	×7

3	3	9	9	5	8	3	0	8	6
×7	×4	×7	×5	×5	×2	×7	×6	×6	×6

9	6	8	7	6	8	7	6	4	0
×7	×7	×9	×7	×3	×2	×8	×8	×4	×2

Check your time and answers. If you missed more than ten problems and/or took longer than five minutes to complete the problems, see your instructor as soon as possible. *Do Not Continue On!* If you missed ten or fewer problems, and completed them all in five minutes or less, study the facts you missed and continue on to **Multiplication of Larger Numbers #1**.

SECTION 1-6 MULTIPLICATION OF LARGER NUMBERS

1.
$$\begin{array}{r} 8250 \\ \times\ 307 \\ \hline 0 \end{array}$$

   Multiply 7 times 0.

$$\begin{array}{r} \overset{3}{}8250 \\ \times\ 307 \\ \hline 50 \end{array}$$

   Multiply 7 times 5 and indicate "carry-over."

$$\begin{array}{r} \overset{13}{}8250 \\ \times\ 307 \\ \hline 750 \end{array}$$

   Multiply 7 times 2 and add the "carry-over" to the product.
   $7 \times 2 = 14$
   $14 + 3 = 17$
   Indicate the "carry-over."

$$\begin{array}{r} \overset{1}{}8250 \\ \times\ 307 \\ \hline 57750 \end{array}$$

   Multiply 7 times 8 and add the "carry-over" to the product.
   $7 \times 8 = 56$
   $56 + 1 = 57$

$$\begin{array}{r} 8250 \\ \times\ 307 \\ \hline 57750 \\ 0000 \end{array}$$

   Multiply 0 times 0
   0 times 5
   0 times 2
   0 times 8

$$\begin{array}{r} \overset{1}{}8250 \\ \times\ 307 \\ \hline 57750 \\ 0000 \\ 50 \end{array}$$

   Multiply 3 times 0.
   Multiply 3 times 5 and indicate "carry-over."

$$\begin{array}{r} \overset{1}{}8250 \\ \times\ 307 \\ \hline 57750 \\ 0000 \\ 24750 \\ \hline 2532750 \end{array}$$

   Multiply 3 times 2 and add "carry-over — $3 \times 2 = 6$
   $6 + 1 = 7$
   Multiply 3 times 8
   Add the columns
   = **Answer**

2.
$$\begin{array}{r}\overset{1}{400\overset{|}{6}}\\ \times\ 12\overset{|}{3}\\ \hline 8\end{array}$$

Multiply 3 × 6 and indicate "carry-over."

$$\begin{array}{r}\overset{1}{400\overset{}{6}}\\ \times\ 123\\ \hline 18\end{array}$$

Multiply 3 × 0 and add "carry-over" to product: 3 × 0 = 0
0 + 1 = 1

$$\begin{array}{r}4006\\ \times\ 123\\ \hline 12018\end{array}$$

Multiply 3 × 0
Multiply 3 × 4

$$\begin{array}{r}\overset{1}{400\overset{}{6}}\\ \times\ 12\overset{}{3}\\ \hline 12018\\ 2\end{array}$$

Multiply 2 × 6 and indicate "carry-over."

$$\begin{array}{r}\overset{1}{400\overset{|}{6}}\\ \times\ 12\overset{}{3}\\ \hline 12018\\ 12\end{array}$$

Multiply 2 × 0 and then add "carry-over:" 2 × 0 = 0
0 + 1 = 1

$$\begin{array}{r}4006\\ \times\ 123\\ \hline 12018\\ 8012\end{array}$$

Multiply 2 × 0
Multiply 2 × 4

$$\begin{array}{r}4006\\ \times\ 123\\ \hline 12018\\ 8012\\ 4006\\ \hline 492738\end{array}$$ = **Answer**

Multiply 1 × 6
Multiply 1 × 0
Multiply 1 × 0
Multiply 1 × 4
Add columns

3.
$$\begin{array}{r}{\overset{4}{7098}}\\ \times\ 76\\ \hline 8\end{array}$$

Multiply 6 × 8 and indicate "carry-over."

$$\begin{array}{r}{\overset{54}{7098}}\\ \times\ 76\\ \hline 88\end{array}$$

Multiply 6 × 9 and add "carry-over:"   6 × 9 = 54
                                                  54 + 4 = 58

Then indicate "carry-over."

$$\begin{array}{r}{\overset{5}{7098}}\\ \times\ 76\\ \hline 42588\end{array}$$

Multiply 6 × 0 and add "carry-over:"   6 × 0 = 0
                                                  0 + 5 = 5

Multiply 6 × 7

$$\begin{array}{r}{\overset{5}{7098}}\\ \times\ 76\\ \hline 42588\\ 6\end{array}$$

Multiply 7 × 8
Indicate "carry-over."

$$\begin{array}{r}{\overset{65}{7098}}\\ \times\ 76\\ \hline 42588\\ 86\end{array}$$

Multiply 7 × 9 and add "carry-over:"   7 × 9 = 63
                                                63 + 5 = 68

Then indicate "carry-over."

$$\begin{array}{r}{\overset{6}{7098}}\\ \times\ 76\\ \hline 42588\\ 49686\\ \hline \underline{539448}\end{array}$$ = **Answer**

Multiply 7 × 0 and add "carry-over:"   7 × 0 = 0
                                                0 + 6 = 6

Multiply 7 × 7
Add columns

Multiplication of Larger Numbers #1

* 12 × 4	22 × 3	34 × 2	23 × 3	11 × 7	24 × 2	21 × 4	21 × 3	21 × 7	* 53 × 3
21 × 8	31 × 6	64 × 2	* 41 × 8	83 × 3	92 × 4	65 × 1	92 × 4	* 71 × 5	30 × 2
43 × 9	56 × 7	72 × 8	65 × 5	* 59 × 4	87 × 3	45 × 2	36 × 3	77 × 5	* 93 × 6
12 × 12	13 × 12	* 23 × 23	33 × 33	42 × 21	61 × 11	24 × 22	44 × 22	* 51 × 16	* 48 × 11
55 × 37	68 × 26	* 77 × 43	59 × 21	87 × 23	75 × 34	38 × 43	17 × 35	* 94 × 24	32 × 49
800 × 29	64 × 27	730 × 42	803 × 26	* 795 × 30	606 × 40	* 300 × 69	832 × 48	672 × 56	* 305 × 47
618 × 34	743 × 25	912 × 18	452 × 16	345 × 32	787 × 60	900 × 48	100 × 31	473 × 60	980 × 70
822 × 43	* 234 × 12	123 × 32	322 × 23	534 × 22	413 × 23	512 × 23	* 2332 × 123	1232 × 321	3113 × 232
6275 × 128	9735 × 849	* 3692 × 463	5649 × 382	7270 × 620	8000 × 600	* 3601 × 851	7592 × 362		

Check your answers. If you missed more than five problems, study the procedure for multiplication of larger numbers by looking at the solutions to the asterisked problems; then proceed to **Multiplication of Larger Numbers #2.** Otherwise, continue on to **Basic Division #1.**

Multiplication of Larger Numbers #2

| 12 × 3 | 33 × 2 | *24 × 2 | 43 × 3 | 11 × 8 | 24 × 2 | 61 × 4 | 31 × 5 | 34 × 2 | *21 × 6 |

| 31 × 7 | 51 × 8 | 63 × 3 | 74 × 2 | 93 × 3 | 92 × 4 | 21 × 5 | *33 × 0 | *40 × 5 | 65 × 1 |

| 34 × 9 | 65 × 7 | 85 × 9 | 63 × 8 | 29 × 8 | 37 × 7 | 62 × 6 | 48 × 3 | 95 × 5 | *108 × 8 |

| 13 ×12 | *13 ×13 | 32 ×32 | 41 ×22 | 51 ×11 | 22 ×24 | *24 ×42 | 23 ×23 | 66 ×11 | 48 ×10 |

| 45 ×23 | 67 ×82 | 29 ×33 | 60 ×49 | *87 ×36 | 25 ×15 | 32 ×46 | 38 ×19 | 75 ×26 | *60 ×70 |

| 809 × 39 | *901 × 83 | 300 ×245 | 650 × 85 | 720 × 39 | *893 × 20 | 647 × 50 | 600 × 30 | 897 × 25 | 397 × 23 |

| 690 × 38 | 201 × 36 | *300 × 18 | 927 × 63 | 452 × 39 | 859 × 47 | 221 × 63 | 154 × 65 | 999 × 99 | *880 × 66 |

| 901 × 43 | 800 × 64 | *624 × 12 | 231 × 32 | 233 × 23 | 543 × 22 | 423 × 13 | 612 × 43 | 6211 × 312 | 4321 × 212 |

| 1500 × 600 | 1398 × 295 | 6700 × 890 | 1583 × 200 | *6709 × 809 | 3000 × 400 | *7050 × 305 | 4598 × 675 | | |

Check your answers. If you missed more than five problems, study the problems you missed, and take **Multiplication of Larger Numbers #3**. Otherwise, continue on to **Basic Division #1**.

Multiplication of Larger Numbers #3

10 × 4	33 × 3	14 × 2	23 × 3	31 × 4	11 × 7	10 × 8	21 × 7	64 × 2	92 × 4
11 × 6	40 × 8	13 × 3	50 × 5	84 × 2	43 × 3	61 × 5	73 × 2	99 × 0	40 × 7
43 × 9	65 × 7	56 × 8	27 × 9	95 × 4	78 × 3	45 × 2	36 × 3	93 × 7	69 × 9
12 × 11	13 × 12	24 × 12	21 × 42	16 × 11	22 × 24	44 × 22	33 × 33	*701 × 11	20 × 14
65 × 15	70 × 83	85 × 97	60 × 32	78 × 42	69 × 50	*80 × 40	30 × 55	79 × 16	22 × 55
800 × 56	910 × 60	370 × 20	865 × 55	937 × 28	679 × 34	892 × 76	345 × 40	*700 × 86	*909 × 39
756 × 43	987 × 32	675 × 88	900 × 37	609 × 50	843 × 70	626 × 12	189 × 77	300 × 48	*707 × 40
125 × 43	234 × 11	132 × 23	323 × 23	534 × 32	314 × 34	215 × 21	231 × 21	323 × 112	521 × 341
1001 × 729	6012 × 381	*9102 × 602	1800 × 910	6220 × 347	8218 × 659	4227 × 983	1675 × 900		

Check your answers. If you have missed more than five problems, see your instructor immediately. Do *not* continue on. If you missed five or fewer problems, proceed to **Basic Division #1.**

SECTION 1-7 BASIC DIVISION

Basic Division #1

Time Yourself

16 ÷ 8 =	32 ÷ 4 =	16 ÷ 2 =	2 ÷ 2 =	4 ÷ 1 =
8 ÷ 1 =	9 ÷ 3 =	15 ÷ 5 =	16 ÷ 4 =	4 ÷ 2 =
10 ÷ 5 =	2 ÷ 1 =	0 ÷ 5 =	8 ÷ 4 =	20 ÷ 5 =
25 ÷ 5 =	21 ÷ 7 =	18 ÷ 3 =	16 ÷ 1 =	24 ÷ 3 =
27 ÷ 9 =	24 ÷ 6 =	30 ÷ 5 =	35 ÷ 7 =	42 ÷ 6 =
45 ÷ 5 =	36 ÷ 9 =	32 ÷ 4 =	49 ÷ 7 =	36 ÷ 6 =
32 ÷ 8 =	35 ÷ 5 =	42 ÷ 7 =	45 ÷ 9 =	56 ÷ 8 =
63 ÷ 9 =	64 ÷ 8 =	72 ÷ 9 =	56 ÷ 7 =	42 ÷ 7 =
0 ÷ 27 =	56 ÷ 8 =	54 ÷ 6 =	81 ÷ 9 =	14 ÷ 7 =
10 ÷ 2 =	36 ÷ 4 =	18 ÷ 9 =	64 ÷ 8 =	72 ÷ 8 =
63 ÷ 7 =	54 ÷ 6 =	0 ÷ 21 =	15 ÷ 5 =	8 ÷ 1 =
81 ÷ 9 =	72 ÷ 9 =	54 ÷ 9 =	48 ÷ 8 =	40 ÷ 5 =
45 ÷ 9 =	49 ÷ 7 =	24 ÷ 3 =	28 ÷ 7 =	32 ÷ 4 =
35 ÷ 7 =	54 ÷ 6 =	48 ÷ 6 =	21 ÷ 3 =	63 ÷ 9 =
81 ÷ 9 =	15 ÷ 5 =	20 ÷ 4 =	25 ÷ 5 =	0 ÷ 63 =
24 ÷ 4 =	18 ÷ 9 =	63 ÷ 9 =	54 ÷ 6 =	56 ÷ 8 =
48 ÷ 8 =	42 ÷ 6 =	49 ÷ 7 =	45 ÷ 9 =	54 ÷ 9 =
72 ÷ 9 =	36 ÷ 6 =	32 ÷ 8 =	0 ÷ 49 =	28 ÷ 7 =
56 ÷ 7 =	48 ÷ 6 =	72 ÷ 9 =	63 ÷ 9 =	56 ÷ 7 =
0 ÷ 9 =	24 ÷ 6 =	35 ÷ 7 =	32 ÷ 4 =	36 ÷ 9 =

Check your *time* and *answers*. If you missed more than ten problems and/or took longer than five minutes to complete them, then study the problems you missed and take **Basic Division #2**. Otherwise, study the facts you missed and continue on to **Division of Larger Numbers #1**.

Basic Division #2

Time Yourself

0 ÷ 5 =	10 ÷ 5 =	2 ÷ 2 =	1 ÷ 1 =	6 ÷ 3 =
4 ÷ 2 =	10 ÷ 2 =	9 ÷ 3 =	12 ÷ 4 =	15 ÷ 5 =
18 ÷ 6 =	24 ÷ 4 =	27 ÷ 9 =	42 ÷ 7 =	45 ÷ 5 =
63 ÷ 9 =	56 ÷ 7 =	12 ÷ 4 =	36 ÷ 9 =	42 ÷ 6 =
54 ÷ 9 =	24 ÷ 8 =	12 ÷ 3 =	48 ÷ 6 =	45 ÷ 9 =
72 ÷ 8 =	63 ÷ 9 =	56 ÷ 8 =	49 ÷ 7 =	15 ÷ 3 =
10 ÷ 2 =	36 ÷ 4 =	28 ÷ 7 =	49 ÷ 7 =	42 ÷ 6 =
27 ÷ 3 =	24 ÷ 6 =	56 ÷ 7 =	63 ÷ 7 =	4 ÷ 4 =
10 ÷ 5 =	18 ÷ 2 =	81 ÷ 9 =	64 ÷ 8 =	48 ÷ 8 =
40 ÷ 5 =	35 ÷ 7 =	32 ÷ 4 =	28 ÷ 4 =	25 ÷ 5 =
35 ÷ 5 =	54 ÷ 6 =	63 ÷ 9 =	72 ÷ 8 =	40 ÷ 5 =
32 ÷ 8 =	49 ÷ 7 =	64 ÷ 8 =	54 ÷ 9 =	16 ÷ 4 =
30 ÷ 5 =	81 ÷ 9 =	54 ÷ 6 =	45 ÷ 9 =	63 ÷ 9 =
81 ÷ 9 =	64 ÷ 8 =	56 ÷ 7 =	36 ÷ 9 =	45 ÷ 9 =
32 ÷ 8 =	10 ÷ 2 =	4 ÷ 4 =	8 ÷ 2 =	9 ÷ 3 =
72 ÷ 9 =	56 ÷ 7 =	48 ÷ 6 =	42 ÷ 7 =	45 ÷ 9 =
12 ÷ 6 =	20 ÷ 5 =	18 ÷ 3 =	24 ÷ 3 =	0 ÷ 9 =
32 ÷ 8 =	48 ÷ 6 =	54 ÷ 9 =	63 ÷ 9 =	72 ÷ 8 =

Check your *time* and your *answers*. If you missed more than ten problems and/or took longer than five minutes to complete them, study the problem you missed and take **Basic Division #3**. Otherwise, study the facts you missed and continue on to **Division of Larger Numbers #1**.

Basic Division #3

Time Yourself

10 ÷ 5 =	4 ÷ 1 =	16 ÷ 2 =	18 ÷ 9 =	25 ÷ 5 =
32 ÷ 4 =	42 ÷ 6 =	35 ÷ 5 =	45 ÷ 9 =	56 ÷ 7 =
12 ÷ 3 =	15 ÷ 3 =	10 ÷ 2 =	16 ÷ 4 =	32 ÷ 8 =
49 ÷ 7 =	63 ÷ 9 =	24 ÷ 4 =	81 ÷ 9 =	36 ÷ 6 =
42 ÷ 7 =	10 ÷ 5 =	12 ÷ 4 =	16 ÷ 8 =	24 ÷ 6 =
16 ÷ 4 =	20 ÷ 5 =	32 ÷ 4 =	36 ÷ 9 =	30 ÷ 6 =
45 ÷ 5 =	54 ÷ 6 =	48 ÷ 6 =	64 ÷ 8 =	0 ÷ 63 =
54 ÷ 6 =	32 ÷ 8 =	16 ÷ 4 =	63 ÷ 7 =	72 ÷ 9 =
72 ÷ 8 =	54 ÷ 9 =	32 ÷ 8 =	36 ÷ 4 =	81 ÷ 9 =
15 ÷ 3 =	16 ÷ 16 =	25 ÷ 5 =	49 ÷ 7 =	64 ÷ 8 =
54 ÷ 6 =	63 ÷ 9 =	48 ÷ 6 =	54 ÷ 6 =	0 ÷ 9 =
24 ÷ 6 =	18 ÷ 6 =	56 ÷ 7 =	48 ÷ 8 =	40 ÷ 5 =
56 ÷ 8 =	48 ÷ 8 =	36 ÷ 6 =	54 ÷ 6 =	0 ÷ 27 =
27 ÷ 9 =	32 ÷ 4 =	42 ÷ 6 =	54 ÷ 9 =	64 ÷ 8 =
81 ÷ 9 =	4 ÷ 4 =	28 ÷ 4 =	36 ÷ 4 =	12 ÷ 2 =
54 ÷ 9 =	56 ÷ 8 =	64 ÷ 8 =	72 ÷ 9 =	56 ÷ 7 =
63 ÷ 9 =	81 ÷ 9 =	0 ÷ 81 =	36 ÷ 6 =	45 ÷ 5 =
56 ÷ 8 =	27 ÷ 9 =	28 ÷ 7 =	30 ÷ 6 =	40 ÷ 8 =

Check your *time* and your *answers*. If you missed more than ten problems and/or took longer than five minutes to complete them, see your instructor as soon as possible. Do *not* continue on! If you missed fewer than ten problems and completed them in less than five minutes, then study the facts you missed and continue on to **Division of Larger Numbers #1.**

SECTION 1-8 DIVISION OF LARGER NUMBERS

1. 6
    $45\overline{)30956}$

    Divide 45 into 309 and place answer *above* the 9 (the last digit divided into).

    6
    $45\overline{)30956}$
    270

    Multiply 6 times 45. Write answer below 309.

    6
    $45\overline{)30956}$
    -270
    39

    Subtract 270 *from* 309.

    $68\leftarrow$
    $45\overline{)30956}$
    $-270\downarrow$
    $\rightarrow 395$

    Bring down the next number (5). Divide 45 into 395 and place answer above the 5 (the last digit divided into).

    68
    $45\overline{)30956}$
    -270
    395
    360

    Multiply 8 × 45. Write the answer below 395.

    68
    $45\overline{)30956}$
    -270
    395
    $-360\downarrow$
    356

    Subtract 360 *from* 395, *then* bring down next number (6).

    $687\leftarrow$
    $45\overline{)30956}$
    -270
    395
    -360
    $\rightarrow 356$

    Divide 45 into 356. Write answer above the 6.

```
      687 R41
 45)30956
  − 270
    395
   − 360
     356
    − 315
      41
```

Multiply 7 times 45 and write answer below 356. Subtract 315 from 356. At this point, since there are no more numbers to bring down, the division is complete. We write 41 in the answer as R41 to indicate a remainder of 41.

Answer is 687 R41

CHECK: Multiply 45 times 687, then add 41 (the remainder) to the answer. If the total is equal to 30956 (the number we divided into) then our answer, 687 R41, is correct.

```
     687
  ×   45
    3435
    2748
   30915
 +    41
   30956    ✓  Checks!
```

2.
```
      3
  13)39546
```
Divide 13 into 39 and place answer *above* the 9.

```
      3
  13)39546
   − 39
      0
```
Multiply 3 times 13 and write answer below 39. Subtract 39 from 39.

```
      30
  13)39546
   − 39
     05
```
Bring down the next number, 5. Divide 13 into 5—since 13 does *not* divide into 5, place a zero in the answer above the 5.

```
      30
  13)39546
   − 39
      05
    − 00
       54
```
Multiply back, 0 times 13, and place answer below 05. Subtract. Bring down next number.

WHOLE NUMBERS REVIEW   33

$$
\begin{array}{r}
304 \\
13\overline{)39546} \\
-39 \\
\hline
05 \\
-00 \\
\hline
54
\end{array}
$$

Divide 13 into 54 and place answer above the 4.

$$
\begin{array}{r}
304 \\
13\overline{)39546} \\
-39 \\
\hline
05 \\
-00 \\
\hline
54 \\
-52 \\
\hline
26
\end{array}
$$

Multiply back, 4 times 13, and place answer below 54.
Subtract and bring down next number.

$$
\begin{array}{r}
3042 \\
13\overline{)39546} \\
-39 \\
\hline
05 \\
-00 \\
\hline
54 \\
-52 \\
\hline
26 \\
-26 \\
\hline
0
\end{array}
$$

Divide 13 into 26 and place answer above 6.
Multiply back, 2 times 13, and place answer below 26.
Subtract.

Answer is 3042

CHECK:

$$
\begin{array}{r}
3042 \\
\times 13 \\
\hline
9126 \\
3042 \\
\hline
39546
\end{array}
$$

✓ *Checks!*

3. $$
\begin{array}{r}
3 \\
257\overline{)8483084}
\end{array}
$$

Divide 257 into 848.

$$
\begin{array}{r}
3 \\
257\overline{)8483084} \\
-771 \\
\hline
773
\end{array}
$$

Multiply 3 times 257.
Subtract.
Bring down next number (3).

34 CHAPTER 1

```
        33←
257)8483084
    −771
      773
     −771↓
        20
```

Divide 257 into 774.
Multiply 3 times 257.
Subtract.
Bring down next number (0).

```
       330←
257)8483084
    −771
      773
     −771
        20
       − 0↓
        208
```

Divide 257 into 20.
Multiply 0 times 257.
Subtract.
Bring down next number (8).

```
      3300←
257)8483084
    −771
      773
     −771
        20
       − 0
        208
       −  0↓
        2084
```

Divide 257 into 208.
Multiply 0 times 257.
Subtract.
Bring down next number (4).

```
     33008←
257)8483084
    −771
      773
     −771
        20
       − 0
        208
       −  0
        2084
       −2056
          28 → Remainder
```

Divide 257 into 2084.
Multiply 8 times 257.
Subtract.

Answer is 33008 R28

CHECK:
```
        33008
      ×   257
       231056
       165040
        66016
      8483056
    +      28
      8483084  ✓ Checks!
```

Division of Larger Numbers #1

4)92   6)96   7)91   5)90   8)96

6)84   7)84   *4)96   4)72   3)84

4)568   8)968   7)847   * 5)750   3)423

8)896   7)784   6)672   5)595   4)896

33)165   43)172   15)75   14)84   13)91

25)150   15)135   23)115   32)128   42)126

12)3660   24)2472   17)3400   * 13)4605   * 20)6100

* 10)60400   15)480   65)650   84)18648   25)126000

56)18144   34)272748   10)6570   30)288600   14)728

*16)16032   64)64128   31)4960   82)66912   98)6370

172)4128   214)7490   *312)9984   413)5369   216)48384

125)4000   242)14762   304)33744   *507)304707   600)4800

9)83   *16)642   32)901   18)546   80)6046

152)7589   645)19080   100)5678   309)8093   1)61958746

Check your answers! If you've missed more than five problems, study the problems you've answered incorrectly (notice the solutions to the asterisk (*) problems in the back of this book) and take **Division of Larger Numbers #2**. Otherwise you are ready to take the **Chapter 1 Review Test** (at the end of this chapter). After completing that review test, you should be ready to ask your instructor for the topic quiz on whole numbers.

Division of Larger Numbers #2

4)100   6)192   7)357   9)108   6)90

5)120   8)240   4)96   7)287   8)280

6)570   7)406   8)512   9)702   *5)250

8)496   4)848   3)855   2)182   6)108

12)48   15)135   22)154   17)153   35)280

57)285   61)366   87)261   93)372   54)324

21)1827   45)4545   64)5248   *55)11385 60)12120

37)3700   41)21361   83)5312   97)9409   54)3618

38 CHAPTER 1

48)28848	*27)13500	90)41220	81)51840	35)1225
17)918	68)3264	*43)8901	92)86572	28)14224
182)17836	675)166725	*301)2710204	586)609440	974)312654
840)272160	600)342000	721)711627	*683)466489	105)21525
6)74	13)740	25)476	83)298	74)3650
10)7235	586)736	690)8945	321)14568	73210)0

Check your answers. If you missed more than five problems, study the problems you answered incorrectly and take **Division of Larger Numbers #3.** Otherwise you are now ready to take the **Chapter 1 Review Test** (at the end of this chapter.) After completing that review test, you should be ready to ask your instructor for the topic quiz on whole numbers.

Division of Larger Numbers #3

5)490   6)192   7)168   8)240   2)42

9)603   8)120   3)285   4)232   6)540

7)595   *4)240   9)729   7)441   3)141

6)468   8)288   7)413   9)306   4)228

13)117   40)160   16)80   28)112   70)490

15)120   34)238   62)434   75)450   94)282

27)2754   58)1856   10)900   *59)1180   43)2451

82)2542   71)2130   *99)9801   40)2520   68)3672

40 CHAPTER 1

96)19584   23)14490   47)3196   83)76692   *19)193819

37)30747   89)55447   74)43438   33)31944   66)529386

529)3174   105)735   300)1500   727)23264   606)7272

155)5735   280)85120   397)79400   426)23856   700)737800

7)58   16)723   28)6370   35)4982   87)6070

100)29385   *403)823   790)6700   385)123501   1)18573659

 Check your answers. If you've missed more than five problems, see your instructor *before* continuing on. Otherwise, take the **Chapter 1 Review Test.** Upon completing that test, you should be ready to ask your instructor for the topic quiz on whole numbers.

CHAPTER 1

REVIEW TEST #1

Perform the indicated operations and keep track of your time.

See Sections 1-1 and 1-2 for examples of these problems.

1.   23
     + 72

2.   85
     + 96

3.   470
     + 36

4.   5000
      463
     + 79

5.   2138
      1796
     + 3175

See Sections 1-3 and 1-4 for examples of these problems.

6.   76
     − 9

7.   37
     − 18

8.   199
     − 37

9.   3000
     − 465

10. 9081
     − 573

See Sections 1-5 and 1-6 for examples of these problems.

11.   32
     × 8

12.   68
     × 49

13. 385
     × 42

14. 496
     × 37

15. 2001
     × 907

See Sections 1-7 and 1-8 for examples of these problems.

16. $16\overline{)320}$

17. $34\overline{)10472}$

18. $60\overline{)5198}$

19. $301\overline{)60340}$

20. $490\overline{)586013}$

CHECK YOUR ANSWERS

If you missed no more than one problem in each operation and took less than half an hour to complete this test, you should be ready to ask your instructor for the Topic Quiz on Whole Numbers. If you missed more than one problem in any operation(s) and/or took longer than half an hour to complete the test, you should correct your mistakes, then take **Chapter 1 Review Test #2**.

CHAPTER 1

REVIEW TEST #2

Perform the indicated operations and keep track of your time.

1. 37
   + 29

2. 68
   + 48

3. 501
   + 99

4. 7011
     688
   + 497

5. 9000
    5173
   + 2872

6. 83
   − 7

7. 57
   − 39

8. 282
   − 65

9. 6005
   − 307

10. 10,000
    − 8792

11. 43
   × 7

12. 27
   × 19

13. 671
   × 89

14. 503
   × 28

15. 7702
   × 430

16. 28)560

17. 71)21584

18. 30)8903

19. 640)8415

20. 305)41050

CHECK YOUR ✓ ANSWERS — If you missed no more than one problem in each operation and took less than half an hour to complete this test, you should be ready to ask your instructor for the Topic Quiz on Whole Numbers. If you missed more than one problem in any operation(s) and/or took longer than half an hour to complete the test, you should correct your mistakes and take **Chapter 1 Review Test #3**.

CHAPTER 1

REVIEW TEST #3

Perform the indicated operations and keep track of your time.

1. 89 + 37

2. 65 + 48

3. 590 + 86

4. 6081 + 437 + 56

5. 3850 + 1234 + 5982

6. 83 − 7

7. 98 − 25

8. 283 − 46

9. 7001 − 860

10. 20,000 − 1765

11. 48 × 7

12. 33 × 24

13. 529 × 56

14. 807 × 37

15. 1008 × 804

16. 39)7800

17. 43)2322

18. 90)180090

19. 608)70736

20. 475)46073

CHECK YOUR ✓ ANSWERS If you missed no more than one problem in each operation and took less than half an hour to complete this test, you should be ready to ask your instructor for the Topic Quiz on Whole Numbers. If you missed more than one problem in any operation(s) and/or took longer than half an hour to complete the test, *see your instructor immediately* for further assistance.

CHAPTER 2: SYMBOLS AND DEFINITIONS

INTRODUCTION

This chapter is designed to introduce you to some of the symbols, terms and basic concepts that you will see used in this book. You should work through this chapter with the idea of becoming *very familiar* with the words and phrases defined here. You do *not,* however, need to memorize this list.

Symbols List

Symbol	Meaning	Example
+	Add	2 + 3 read (2 plus 3)
−	Subtract	4 − 2 read (4 minus 2)
×	Multiply	8 × 3 read (8 times 3)
÷	Divide	10 ÷ 2 read (10 divided by 2)
•	Multiply (raised dot, not a decimal point)	3 • 5 read (3 times 5)
=	Equals	4 = 2 × 2 read (4 equals 2 times 2)
≠	Does not equal	3 ≠ 4 + 1 read (3 does not equal 4 plus 1)
$\stackrel{?}{=}$	Possibly equals or does it equal	$8 + 2 \stackrel{?}{=} 2 + 8$ read (Does 8 + 2 equal 2 + 8)
≅	Approximately equals	19.9 ≅ 20 read (19.9 is approximately equal to 20)

Symbol	Meaning	Example
∴	Therefore	$3 = 2 + 1 \therefore 2 + 1 = 3$
<	Less than	$2 < 3$ read (2 is less than 3)
≤	Less than or equal to	$6 \leq x$ read (6 is less than or equal to x)
>	Greater than	$12 > 9$ read (12 is greater than 9)
≥	Greater than or equal to	$19 \geq x$ read (19 is greater than or equal to x)
( )	Parentheses; grouping symbol	$(8 + 2)$ read (the quantity 8 plus 2)
( )( )	Multiply	$(6)(3)$ means 6 times 3
π	Symbol used to represent the ratio of the circumference of a circle to its diameter. A constant term $\cong 3.1416$ (constant means that the symbol π will always represent the same value of 3.141592 . . .)	$C = \pi d$ read (circumference of a circle equals pi times the diameter of a circle)

SYMBOLS AND DEFINITIONS   **47**

SECTION 2-1 WHOLE AND DECIMAL NUMBERS AND PLACE VALUE

Term	Definition	Example
Decimal System	Our system of writing numbers using the ten digits.	0, 1, 2, 3, 4, 5, 6, 7, 8, 9
Whole Numbers (Integers)	The "counting" numbers. We obtain them by using the ten digits written above.	0, 1, 2, 3, . . . 14, 15, 16 . . . 27, 28, 29 . . .
Place Value	Value of each digit in a whole number.	42: the 4 is in the tens place and ∴ has a value of 4 tens. The 2 is in the ones place and has a value of 2 ones.

Memorize the Place Value Chart.

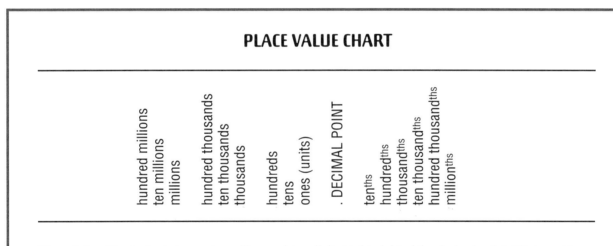

Place Value Chart: A whole number will never have digits to the right of the decimal point. What would you call a number that does have digits to the right of the decimal point?

Term	Definition	Example
Decimal Number or Decimal Fraction	A number which has digits to the right of the decimal point (this definition will be expanded in the chapter dealing with decimals).	GIVEN: 21,094. FIND: (a) Is this a whole number or a decimal number? (b) What are the place values of the digits?

48 CHAPTER 2

Term	Definition	Example

        SOLUTION:
(a) Whole number. Keep in mind that a *whole number* does not have to show its decimal point. It's always assumed to be at the end of a whole number.
(b) From left to right:
    2 ten thousands
    1 thousand
    0 hundreds
    9 tens
    4 ones

GIVEN: 872.6904
FIND: (a) Whole or decimal number?
       (b) Place values?
SOLUTION:
(a) Decimal number—there are digits to the right of the decimal point.
(b) Left to right:
    8 hundreds
    7 tens
    2 ones
    . (decimal point)
    6 ten*ths*
    9 hundred*ths*
    0 thousand*ths*
    4 ten thousand*ths*

GIVEN: 1,100,100
FIND: (a) Whole or decimal number?
       (b) Place values?
SOLUTION:
(a) Whole number—there are no digits to the right of the decimal point.
    1,100,100 = 1,100,100.
(b) Left to right:
    1 million
    1 hundred thousand
    0 ten thousands
    0 thousands
    1 hundred
    0 tens
    0 ones

EXERCISE 2-1

In the following problems, find:
a. Is the number given a whole number or a decimal number?
b. Give the place values of the digits in the given number.

1. 2
2. 8
3. 736.
4. 1.095
5. 1,238,543
6. 83.96
7. 100.0097
8. .31528
9. 15,000.
10. 235,576.9
11. 18.5
12. 36
13. 3,600
14. 150.98
15. 8. 3652
16. 4,513.
17. 800
18. 953,678.236
19. 4,101,370
20. 111,111,111.111111

SECTION 2-2 WORDS USED FOR MATHEMATICAL OPERATIONS

Term	Definition	Example
Sum	Answer to an addition problem.	Find the *sum* of 28 and 42 means to *add* 28 and 42 together. $28 + 42 = 70$; also $\begin{array}{r} 28 \\ + 42 \\ \hline 70 \end{array}$ ∴ 70 is the sum of 28 and 42. Find the sum of 2, 6, 7 and 1 means: $2 + 6 + 7 + 1 = 16$; also $\begin{array}{r} 2 \\ 6 \\ 7 \\ + 1 \\ \hline 16 \end{array}$ ∴ 16 is the sum of 2, 6, 7 and 1.
Difference	Answer to a subtraction problem.	Find the *difference* of 12 and 9 means to *subtract* 9 from 12: $12 - 9 = 3$; also $\begin{array}{r} 12 \\ - 9 \\ \hline 3 \end{array}$ ∴ 3 is the difference of 12 and 9.
Product	Answer to a multiplication problem.	Find the product of 6 and 10 means to *multiply* 6 by 10: $6 \times 10 = 60$ or $6 \bullet 10 = 60$ ∴ 60 is the product of 6 and 10. Find the product of 2, 3 and 4 means: $2 \times 3 \times 4 = 24$; or $2 \bullet 3 \bullet 4 = 24$; or $(2)(3)(4) = 24$ ∴ 24 is the product of 2, 3 and 4.
Quotient	Answer to a division problem.	Find the *quotient* of 54 and 9 means to *divide* 54 by 9: $54 \div 9 = 6$ or $9\overline{)54}\,^{6}$ or $\dfrac{54}{9} = 6$ ∴ 6 is the quotient of 54 and 9.

EXERCISE 2-2

In this exercise, rewrite the given problem using the appropriate operation symbol and solve.

Example: **Find the product of 5 and 2.**   *Solution:* **5 × 2 = 10**

1. Find the sum of 3 and 6.

2. Find the difference of 8 and 3.

3. Find the product of 9 and 2.

4. Find the quotient of 25 and 5.

5. Find the difference of 10 and 1.

6. Find the quotient of 36 and 6.

7. Find the product of 20 and 2.

8. Find the sum of 13, 8 and 4.

9. Find the product of 5, 5 and 2.

10. Find the sum of 8, 9, 12 and 3.

11. Find the quotient of 0 and 3.

12. Find the product of 0 and 2.

13. Find the difference of 13 and 13.

14. Find the sum of 0 and 4.

15. Find the quotient of 8 and 8.

SECTION 2-3 FACTORS, PRIME AND COMPOSITE NUMBERS

Term	Definition	Example
Factors	Whole numbers which divide evenly into other whole numbers, or, in other words, give quotients that are whole numbers (no remainder).	Factors of 10 are 1, 2, 5, 10. 1, 2, 5 and 10 will all divide into 10 leaving no remainder. Factors of 11 are 1, and 11. 1 and 11 will divide evenly into 1.
Prime number	A whole number that has *only two factors:* itself and 1. It cannot be evenly divided by anything other than itself and 1.	2 is a prime number since the only factors of 2 are 2 and 1. 3 is a prime number. 5 is a prime number. 7 is a prime number.
Composite Number	A whole number which has *more than two factors.* A composite number is a whole number that is not a prime number.	4 is a composite number since its factors are 1, 2 and 4. 6 is a composite number; 1, 2, 3 and 6 are its factors. 8 is a composite number; 1, 2, 4, and 8 are its factors.

EXERCISE 2-3

(Read very carefully!)

1. Is 4 a factor of 8?
2. Is 6 a factor of 18?
3. Is 3 a factor of 10?
4. Is 9 a factor of 3?
5. Is 2 a factor of 12?
6. Is 10 a prime or composite number?
7. Is 11 a prime or composite number?
8. Is 9 a prime or composite number?
9. Is 21 a prime or composite number?
10. Is 17 a prime or composite number?

SECTION 2-4 EXPONENTIAL NOTATION

Term	Definition	Example
Exponential Notation	"Shorthand" method of writing repeated multiplication of the *same* factor.	$2 \cdot 2 \cdot 2 \cdot 2$ can be written 2^4. ← exponent The exponent refers to the number of times the factor occurs in the multiplication. 2^4 is exponential notation. $3 \cdot 3 = 3^2$ $8 \cdot 8 \cdot 8 \cdot 8 \cdot 8 \cdot 8 \cdot 8 = 8^7$ $4 \cdot 4 \cdot 4 \cdot 3 \cdot 3 \cdot 2 \cdot 2 \cdot 2 \cdot 2 = 4^3 \cdot 3^2 \cdot 2^4$

How to Read Numbers in Exponential Notation

$5 = 5^1$ — This is read "five to the first power." Usually the exponent 1 is not written, but it is "assumed" to be there.

$5 \cdot 5 = 5^2$ — This is read "five to the second power" or more commonly "five squared."

$5 \cdot 5 \cdot 5 = 5^3$ — This is read "five to the third power" or more commonly "five cubed."

$5 \cdot 5 \cdot 5 \cdot 5 = 5^4$ — This is read "five to the fourth power."

$5 \cdot 5 \cdot 5 \cdot 5 \cdot 5 = 5^5$ — This is read "five to the fifth power."

And so on . . .

How would you read 3^4? If you said "three to the fourth power" you're right! If you didn't, read through "How to read numbers in exponential notation" again.

Simplifying a Problem Containing Exponents

To simplify problems containing exponential notation, write each factor as many times as indicated by the exponent, carry out the multiplication of those factors, then perform the indicated operation.

EXAMPLE *Simplify*
$$2^3 + 4^2 = 2 \cdot 2 \cdot 2 + 4 \cdot 4$$
$$= 8 + 16$$
$$= 24$$

EXAMPLE *Simplify*
$$3^2 + 5^2 + 2^3 = 3 \cdot 3 + 5 \cdot 5 + 2 \cdot 2 \cdot 2$$
$$= 9 + 25 + 8$$
$$= 42$$

EXAMPLE   Simplify $3^4 \cdot 5^2 = 3 \cdot 3 \cdot 3 \cdot 3 \cdot 5 \cdot 5$
   $= 81 \cdot 25$
   $= 2025$

EXAMPLE   Simplify $2^3 \cdot 5 \cdot 4^2 = 2 \cdot 2 \cdot 2 \cdot 5 \cdot 4 \cdot 4$
   $= 8 \cdot 5 \cdot 16$
   $= 640$

EXERCISE 2-4   *Make sure you work ALL of this exercise.*

In problems 1–10 write the given factors in exponential notation.

1. $3 \cdot 3 \cdot 3 \cdot 3$
2. $5 \cdot 5$
3. $6 \cdot 6 \cdot 6$
4. $10 \cdot 10 \cdot 10 \cdot 10 \cdot 10 \cdot 10$
5. $100 \cdot 100 \cdot 100$
6. $2 \cdot 2 \cdot 3 \cdot 3 \cdot 3 \cdot 4$
7. $5 \cdot 5 \cdot 5 \cdot 2 \cdot 2 \cdot 2 \cdot 3 \cdot 3$
8. $6 \cdot 10 \cdot 10 \cdot 10 \cdot 10$
9. $8 \cdot 8 \cdot 9 \cdot 9 \cdot 9 \cdot 15 \cdot 15$
10. $2 \cdot 2 \cdot 5 \cdot 5 \cdot 5 \cdot 11 \cdot 11 \cdot 3$

Simplify problems 11–20.

*11. $2^2 + 4^2$
12. $3^3 + 2^3$
13. $10^2 + 3^3$
*14. $4^3 \cdot 2^4$
15. $2^5 \cdot 3$
*16. $3^2 + 2^3 + 5^2$
17. $5 + 6^2 + 7^2$
18. $2^3 \cdot 4^2 \cdot 5^3$
19. $6^2 \cdot 7^2 \cdot 5$
*20. $8^2 \cdot 2^3 \cdot 4^2$

SECTION 2-5 ORDER OF OPERATIONS

Term	Definition	Example
Order of Operations	Order in which we perform different operations occuring in the same problem.	$3 + 8 \bullet 2 - 6 \div 2 = ?$ Without an "order of operations" we will end up with different answers to the problem, depending on which operations we perform first. There is a need, therefore, for an agreement as to which operations are done first.

ORDER OF OPERATIONS AGREEMENT

STEP 1 Perform operations inside grouping symbols.

STEP 2 Simplify numbers in exponential notation.

STEP 3 Do division and/or multiplication as they occur in order from *left to right*.

STEP 4 Do subtraction and/or addition as they occur in order from *left to right*.

If a problem does not contain one or more of the above conditions (no grouping symbols, no exponents, etc.) then proceed to the next step.

Study the following examples *very carefully*.

EXAMPLE Simplify the following expression using order of operations.

$$8 + 3 \times 2^2 - (6 + 1) \div 7$$

Solution

$8 + 3 \times 2^2 - (6 + 1) \div 7$	STEP 1
$8 + 3 \times 2^2 - (7) \div 7$	STEP 2
$8 + 3 \times 4 - (7) \div 7$	STEP 3
$8 + 12 - 1$	STEP 4
$20 - 1$	
$19 =$ **Answer**	

EXAMPLE *Simplify* $2^2 \times 3 \div (8 - 2) + 5^2$

Solution

$2^2 \times 3 \div (8 - 2) + 5^2$ **STEP 1**

$2^2 \times 3 \div (6) + 5^2$ **STEP 2**

$4 \times 3 \div (6) + 25$ **STEP 3**

$12 \div (6) + 25$

$2 + 25$ **STEP 4**

$27 =$ **Answer**

EXAMPLE *Simplify* $14 \bullet 2 + 4 \div 2 - 8 \bullet 2 + 6$

Solution

$14 \bullet 2 + 4 \div 2 - 8 \bullet 2 + 6$ **STEP 3**

$28 + 2 - 16 + 6$ **STEP 4**

$30 - 16 + 6$

$14 + 6$

$20 =$ **Answer**

EXAMPLE *Simplify* $8^2 - (2^2 + 1) \div (3 + 1^2 + 1^4) + 10 \div 2$

Solution

$8^2 - (2^2 + 1) \div (3 + 1^2 + 1^4) + 10 \div 2$ **STEP 1**

$8^2 - (4 + 1) \div (3 + 1 + 1) + 10 \div 2$

$8^2 - (5) \div (5) + 10 \div 2$ **STEP 2**

$64 - (5) \div (5) + 10 \div 2$ **STEP 3**

$64 - 1 + 5$ **STEP 4**

$63 + 5$

$68 =$ **Answer**

EXAMPLE *Simplify* $\quad 2^3 \cdot 3 \div (4 \div 4) \cdot (24 - 22) - 1 + 16 - 8 \div 2 \cdot 0$

Solution

$$2^3 \cdot 3 \div (4 \div 4) \cdot (24 - 22) - 1 + 16 - 8 \div 2 \cdot 0$$

$$2^3 \cdot 3 \div \quad 1 \quad \cdot \quad 2 \quad - 1 + 16 - 8 \div 2 \cdot 0$$

$$8 \cdot 3 \div \quad 1 \quad \cdot \quad 2 \quad - 1 + 16 - 8 \div 2 \cdot 0$$

$$24$$

$$4$$

$$24$$

$$48 \qquad - 1 + 16 - \qquad 0$$

$$47$$

$$63$$

$$63 = \textbf{Answer}$$

EXAMPLE *Simplify* $\quad 1^5 + 1^7 - 1^8 \cdot 1^{16} + 1^{20} \cdot (2 - 1^4)$

Solution

$$1^5 + 1^7 - 1^8 \cdot 1^{16} + 1^{20} \cdot (2 - 1^4)$$

$$1^5 + 1^7 - 1^8 \cdot 1^{16} + 1^{20} \cdot (2 - 1)$$

$$1^5 + 1^7 - 1^8 \cdot 1^{16} + 1^{20} \cdot \quad 1$$

$$1 + 1 - 1 \cdot 1 + 1 \cdot 1$$

$$1 + 1 - \quad 1 \quad + \quad 1$$

$$2$$

$$1$$

$$2 = \textbf{Answer}$$

EXAMPLE *Simplify* $\quad 40 \div (15 - 13 + 8) \bullet 6 \div 3 + (2^4 + 2^5 - 2 \bullet 2^3) - 3 + 7$

Solution

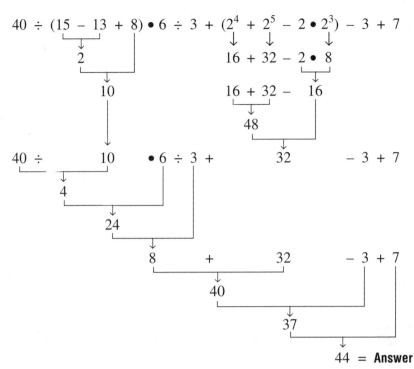

NOTE: In order for a parenthesis to be considered a grouping symbol, there must be *at least one operation* inside a parenthesis.

EXAMPLE $\quad (4 + 8) \quad$ *is* an example of a grouping symbol.
$\qquad\qquad (3) \quad$ is ***not*** an example of a grouping symbol.

EXERCISE 2-5

Simplify problems 1–20 by using order of operations.

1. $6 \div 2 + 1$
2. $8 - 3 + 6$
3. $2 \bullet 3 - 2$
*4. $5 + 3 \bullet 6$
5. $24 - 8 \bullet 2 + 6 \div 2$
6. $3 \bullet 7 - 6 \div 2 + 3 \bullet 8 - 4$
*7. $5 \bullet (2 - 1) \div 5 + 2^2 \bullet (3 + 5)$
8. $6 \div 2 \bullet 3 + 15 - 2 \bullet (4 - 1) \bullet 2^2$
9. $8 - (8 - 2) \div 3 + 3$
10. $2^3 \bullet 3^2 \div 4 \bullet 2$

*11. $18 - 9 \div 3 + 10^2 \bullet 5 \div 2$
12. $(16 - 9) + 3 \bullet 4^2 - (2 + 8) \div 5$
13. $2 + 3 \div 3 + 2$
14. $5 \bullet (2 + 1) + 3^2 - 2 \bullet (5 - 3)$
15. $13 + 6 - 2 \bullet 8 + 7 - 6$
*16. $(18 \div 9)(3 \bullet 6) \div (4)(2)$
17. $4^2 \bullet 2 \bullet (3 \div 3) \div (20 - 4) - 1$
18. $34 - 8 \bullet (2) \div 2 (4) - 1^2$
19. $1^3 - 1^4 + 1^5 + 1^7 \bullet 1^{16} \bullet (1 + 1^2)$
20. $(4 + 3) \div (11 - 2^2) - (1^4)(1)$

SECTION 2-6 COMMUTATIVE AND ASSOCIATIVE PROPERTIES

Term	Definition	Example	
Commutative Property of Addition	Numbers can be *added* in any order and the *sum* will be the same.	2 + 1 + 3 = 6 1 + 2 + 3 = 6 3 + 2 + 1 = 6	Notice: There are the same numbers in each problem, but they have travelled or changed places (commuted).
Commutative Property of Multiplication	Numbers can be *multiplied* in any order and the *product* will be the same.	3 • 4 • 2 = 24 2 • 3 • 4 = 24 4 • 2 • 3 = 24	Notice: Again the numbers are the same in each problem, they have simply commuted or changed places.

The commutative property does *not* hold for subtraction or division.

EXAMPLE 4 ÷ 2 = 2
2 ÷ 4 ≠ 2; 2 ÷ 4 = **.5**   (Not the same answer.)

EXAMPLE 9 − 3 = 6
3 − 9 ≠ 6; 3 − 9 = **−6**   (Not the same answer.)

 The order changed *and* so did the *quotient* and *difference*. As can be seen by the above examples, the commutative property does *not* hold true for division and subtraction. The numbers *cannot* be commuted and still provide the same answer.

EXAMPLES   Determine whether the following are true or false. If true, what property is being applied?

1. 7 + 6 + 5 = 6 + 7 + 5 **Answer:** True—Commutative Property of Addition
2. (16 • 8) • 4 = 4 • (8 • 16) **Answer:** True—Commutative Property of Multiplication
3. 9 ÷ 5 ÷ 4 = 4 ÷ 9 ÷ 5 **Answer:** False
4. 25 + (16 + 3) = (3 + 16) + 25 **Answer:** True—Commutative Property of Addition
5. 3 • 17 • 8 = 17 • 3 • 8 **Answer:** True—Commutative Property of Multiplication

Term	Definition	Example
Associative Property of Addition	Numbers being *added* together can be *grouped* in any order and still give the same result.	(3 + 2) + 5 = (5) + 5 = 10 3 + (2 + 5) = 3 + (7) = 10 Notice the numbers in problems 1 and 2 above did *not* commute (they appear in the same order from left to right); but they are grouped (being associated) differently, therefore changing the order in which the addition is done. The result is the same, however.
Associative Property of Multiplication	Numbers being *multiplied* together can be *grouped* differently and still give the same result.	(8 × 2) × 3 = (16) × 3 = 48 8 × (2 × 3) = 8 × (6) = 48 Notice again, the numbers did not commute, but the order in which the multiplication is done has been changed due to the changed grouping. The numbers are being associated differently, but the result is the same.

The associative property does *not* hold for subtraction or division.

EXAMPLE (3 − 2) − 1 = (1) − 1 = **0**
3 − (2 − 1) = 3 − (1) = **2**   (Not the same answer.)

EXAMPLE (16 ÷ 8) ÷ 2 = (2) ÷ 2 = **1**
16 ÷ (8 ÷ 2) = 16 ÷ 4 = **4**   (Not the same answer.)

NOTE The grouping changed *and* so did the *difference* and *quotient*. The associative property does *not* hold true for subtraction and division; the numbers *cannot* be associated differently and still give the same answer.

EXAMPLES Determine whether the following are true or false. If true, what property is being applied?

1. (7 + 9) + 20 = 7 + (9 + 20)   **Answer:** True—Associative Property of Addition
2. 13 • (2 • 4) = (13 • 2) • 4   **Answer:** True—Associative Property of Multiplication
3.** (8 + 11) + 2 = 2 + (11 + 8)   **Answer:** True—Commutative Property of Addition**

** **NOTICE:** In Example #3, the association of 8 and 11 did *not* change; but the order of the numbers did, therefore, *Commutative* Property of Addition.

EXERCISE 2-6

*Determine whether the following statements are true or false. If **true**, state which property is being applied.*

1. $8 \cdot 3 \cdot 2 = 3 \cdot 8 \cdot 2$ Comm. Mult

2. $4 - 6 - 5 = 6 - 4 - 5$

3. $(14 + 8) + 6 = 14 + (8 + 6)$

4. $(9 \cdot 2) \cdot 2 = 2 \cdot (2 \cdot 9)$ Comm Mult

5. $6 + (3 + 10) = (6 + 3) + 10$ Assoc Add

6. $(4 \div 2) \div 3 = 2 \div (4 \div 3)$

7. $8 \cdot 2 = 2 \cdot 8$

8. $6 + (9 + 7) = (6 + 9) + 7$ Asst Add.

9. $3 \cdot (9 \cdot 2) = (3 \cdot 9) \cdot 2$ Assoc. Mult.

10. $7 + 3 + 2 = 7 + 2 + 3$

11. $(10 \cdot 3) \cdot 6 = 6 \cdot (3 \cdot 10)$

12. $24 + (6 + 13) = 24 + (13 + 6)$

13. $18 \cdot (12 \cdot 11) = (18 \cdot 12) \cdot 11$

14. $21 + 15 + 32 = 32 + 21 + 15$

15. $7 \cdot (18 \cdot 3) = (18 \cdot 3) \cdot 7$

SECTION 2-7 RATIONAL NUMBERS

Term	Definition	Example
Ratio	Comparison of two numbers by division. Ratios can be expressed using a fraction line (/) or a colon (:)	The comparison of π and 2 is the ratio $\pi/2$ (read—pi over two or pi divided by 2) or $\pi:2$ (read the ratio of pi to two)
		The comparison of 5 and 7 is expressed as the ratio 5/7 (read—five-sevenths) or 5:7 (read—the ratio of five to seven)
		The comparison of 8 and 3 is expressed as the ratio 8/3 (read—eight-thirds) or 8:3 (read—the ratio of eight to three)
Rational Numbers	Numbers that can be expressed as the *division of* two *whole* numbers; or in other words, the *ratio* of two *whole* numbers NOTICE: *Ratio*nal	5/8 is a rational number since 5 and 8 are both whole numbers
		2/4 is a rational number since 2 and 4 are both whole numbers
		6 is a rational number since it can be expressed as the ratio 6/1 (which is the ratio of two whole numbers, 6 and 1)

> **NOTE:** All whole numbers are rational numbers since all whole numbers can be expressed as the ratio of itself to one.

Irrational Numbers	Numbers that *cannot* be expressed as the division or ratio of two *whole numbers*	$\sqrt{2}/5$ (read—the square root of two divided by five) is *not* the division of two *whole numbers* ($\sqrt{2}$ is not a whole number) $\therefore$ $\sqrt{2}/5$ is an irrational number
		$\pi/8$ (read—pi divided by eight) is the ratio of two numbers, that are not both whole numbers (π is not a whole number) $\therefore$ $\pi/8$ is an irrational number

> **NOTE:** Most of the numbers that we use and see on a daily basis are rational numbers.

EXERCISE 2-7

State whether the following are rational or irrational numbers and why.

1. 8/9
2. 3/4
3. 5
4. 1/2
5. 100/7
6. 38
7. $\pi/4$
8. $4/\pi$
9. $\sqrt{3}/2$
10. $\sqrt{5}$
11. 3/40
12. 296/82
13. $\sqrt{7}/\sqrt{6}$
14. 65
15. 1
16. $\sqrt{11}/4$
17. 3986
18. 45/67
19. $\pi/8$
20. π

SECTION 2-8 PROCEDURE FOR SOLVING APPLICATION PROBLEMS

Don't let words frighten you! Words are our very good friends. Without them we couldn't speak and have anyone understand us. We also couldn't write or read what we've written. So words are nothing to be afraid of and neither are word problems, which we will refer to from now on as *Applications*. All we are going to do in application problems is take words ("key" words) and turn them into mathematics symbols. Then we'll solve the problem doing exactly the same arithmetic that we've always done.

To help you do all this, follow these five steps and remember: the more familiar you are with doing applications, the easier they become.

STEP 1 Read the problem. Don't try to solve it, just read it to read it. Then read it again. And *read it again*, but this third time try to pick out some important words or "key" words, such as: sum, difference, times, etc.

STEP 2 Very important!! *What are you being asked to find?* If you don't know this you can't solve the problem. So if you have to, read the problem a fourth time to find out what you're solving for.

STEP 3 What information have you been given. This is usually going to be numbers or figures or data. Also at this point you should see a way of getting your answer; either by adding, subtracting, multiplying, or dividing the data; or a combination of these operations.

STEP 4 Once you've determined how you're going to solve the problem, go ahead and perform the necessary operations to get the answer.

STEP 5 Summarize your answer. Make sure you place the proper *units* with the numbers.

Examples of Applications

GIVEN John has four tests to take this week, Judy has two, and Bob has three tests to take this week. Between the three of them, what's the total number of tests that John, Judy, and Bob will take this week?

Solution

STEP 1 Read the problem again *and* again.

STEP 2 What are we being asked to find? We've been asked to find a *total* number of tests.

STEP 3 Given information is: four tests (John), two tests (Judy) and three tests (Bob). If we add these numbers together we should be able to find the total number of tests.

STEP 4 Do it! $4 + 2 + 3 = \boxed{9}$

STEP 5 Summarize: *Nine tests* is the total number of tests John, Judy, and Bob will take this week.

SYMBOLS AND DEFINITIONS   **65**

GIVEN   In one electrical circuit there are two resistors in series and five resistors in parallel. What is the total number of resistors in the circuit?

Solution   **STEP 1**   Read the problem for a second and third time.

STEP 2   We've been asked to find the total number of resistors in the electrical circuit.

STEP 3   The given information is: two resistors (in series) plus five resistors (in parallel). Since we've been asked to find the *total,* we will add 2 plus 5 together.

STEP 4   Do it!   2 + 5 =   7 resistors

STEP 5   Summary: There is a total of *seven resistors* in the circuit.

GIVEN   Usually John works a 40-hour week. However, this week he was sick one day and missed eight hours of work because of it. How many hours did he work this week? (What is the total number of hours he actually worked this week?)

Solution   **STEP 1**   Read.

STEP 2   Number of hours worked this week?

STEP 3   Works 40 hours usually but *missed* eight hours ∴ 40 − 8 = actual number of hours worked this week.

STEP 4   40 − 8 =   32 hours

STEP 5   John worked *32 hours* this week.

GIVEN   A patient is to be given two tablets of medication three times a day. At the end of one day, how many tablets will the patient have received?

Solution   **STEP 1**   Read.

STEP 2   Number of tablets received in one day?

STEP 3   Two tablets given three *times* a day; ∴ 2 × 3 = number of tablets in one day.

STEP 4   2 × 3 =   6 tablets

STEP 5   The patient receives *six tablets* in one day.

GIVEN If ten calculators cost a total of $120, how much does one calculator cost?

Solution

STEP 1 Read.

STEP 2 What does one calculator cost?

STEP 3 Ten calculators cost $120 ∴ $120 ÷ 10 = cost of one calculator.

STEP 4 $120 ÷ 10 = $\boxed{\$12}$

STEP 5 One calculator costs *$12*.

GIVEN Starting with a $100 checking account balance, in one day you write a check for $10; another for $15; and another for $32. In the same day you deposit a check for $50. What is your balance at the end of the day?

Solution

STEP 1 Read.

STEP 2 Checking account balance at end of day?

STEP 3 Write checks for:
$10	Adding these should give
$15	total money to be subtracted
$32	from account.

Starting balance of $100.
Deposit (add to account) $50.

By determining the total amount of money gone from the account, through writing checks, and subtracting that amount from the original balance, then adding in the deposit, we should be able to find our balance at the end of the day.

STEP 4

$10 Total amount of money gone from the account.
 15
 32
———
$57 to be subtracted from $100

$100
− 57
———
$ 43 = balance after writing checks and before deposit of $50.

$43
+ 50 (deposit)
———
$\boxed{\$93}$ = balance at end of day.

STEP 5 At the end of the day there is a *$93* balance in the checking account.

EXERCISE 2-8

1. Ann purchased three books: an algebra text at $105; an economics text at $85; and a psychology text at $56. How much money did she spend all together?

2. A bottle that contains 10,000 U (units) of medication when full now contains 2,000 U. How much medication has been used out of this bottle?

*3. A beam has three concentrated loads placed on it, each load equaling 2k. What is the total load on the beam in kips? (k = kip = *1 kip = 1000 pounds* ∴ 2k = 2 × 1000 pounds = 2000 pounds.)

4. If it takes 20 cups of flour to make four loaves of bread, how much flour is needed for one loaf?

5. In February, Jack worked two 45-hour weeks and two 39-hour weeks. How many hours did he work in February?

6. You are being given a timed quiz. You have 30 minutes to work ten problems. How much time should you allow yourself to work each problem?

*7. If you can drive 60 miles in two hours, how many miles can you drive in five hours? (**HINT:** How many miles can you drive in one hour?)

*8. A new hybrid car is supposed to get 55 miles to the gallon on the highway and 40 miles to the gallon in the city. How many gallons of fuel would be used driving 220 miles on the highway and 160 miles in the city?

9. In problem #8, if the maximum fuel capacity of the car is 10 gallons, how many gallons of fuel are left after driving 220 miles-highway *and* 160 miles-city?

*10. An electronic flash unit on a camera takes ten seconds to charge up to full capacity such that it can be fired. What is the most number of flash pictures you can take in two minutes? (**HINT:** Make sure you work in all *seconds* or all minutes.)

CHAPTER 2

REVIEW TEST #1

1. Give the place values of the digits in the given number:
   a. 149,224 (see 2-1)
   b. 2.0938 (see 2-1)

2. Rewrite the given problem using the appropriate operation symbol and solve:
   a. Find the product of 18 and 36. (see 2-2)
   b. Find the difference of 29 and 2. (see 2-2)

3. Is 5 a factor of 100? (see 2-3)

4. Is 2 a prime number or composite number? (see 2-3)

5. Write the given factors in exponential notation:
   a. $2 \cdot 2 \cdot 5 \cdot 6 \cdot 6 \cdot 6$ (see 2-4)
   b. Simplify: $4^2 + 3^3$ (see 2-4)

6. Simplify using order of operations:
   a. $3 \cdot 4 \div 2 + 6 \cdot 2 - 7$ (see 2-5)
   b. $(8 + 7) \div 3 - 2^2$ (see 2-5)

7. Determine whether the following statement is true or false. If true, what property is being applied?
   a. $6 \cdot 9 = 9 \cdot 6$ (see 2-6)
   b. $(4 + 3) + 5 = 4 + (3 + 5)$ (see 2-6)

8. Are the following rational or irrational numbers:
   a. $\dfrac{4}{3}$ (see 2-7)
   b. $\dfrac{\pi}{8}$ (see 2-7)

9. In March, Beth took a two week paid vacation. If she makes $5 an hour and normally works 40 hours per week, how much money did she make during her vacation? (see 2-8)

CHECK YOUR ANSWERS If you missed more than three (3) answers, correct your mistakes and try Review Test #2. If you answered 12 or more questions correctly, then you should be ready to ask your instructor for the topic quiz on Symbols and Definitions.

CHAPTER 2

REVIEW TEST #2

1. Give the place values of the digits in the given number:
   a. 695
   b. .987

2. Rewrite the given problem using the appropriate operation symbol and solve:
   a. Find the quotient of 9 and 9.
   b. Find the sum of 46 and 23.

3. Is 3 a factor of 12?

4. Is 3 a prime or composite number?

5. Write the given factors in exponential notation:
   a. $10 \cdot 10 \cdot 10 \cdot 15 \cdot 15$
   b. Simplify: $2^3 + 3^2$

6. Simplify using order of operations:
   a. $5 + 8 \cdot 2 - 10 + 3 \div 1^2$
   b. $13 + (9 + 3) - 8 - 3^2 + 2 \cdot 2^3$

7. Determine whether the following statement is true or false. If true, what property is being applied?
   a. $(2 + 5) = (5 + 2)$
   b. $7 + (3 + 2) = (7 + 3) + 2$

8. Are the following rational or irrational numbers?
   a. $\dfrac{5}{\sqrt{3}}$
   b. $\dfrac{3}{5}$

9. If it takes 20 minutes to walk one mile, how many miles can you walk in 140 minutes?

CHECK YOUR ANSWERS — If you missed more than three (3) answers, correct your mistakes and try Review Test #3. If you answered 12 or more questions correctly, then you should be ready to ask your instructor for the topic quiz on Symbols and Definitions.

CHAPTER 2

REVIEW TEST #3

1. Give the place values of the digits in the given number:
   a. 80,970
   b. 2.0536

2. Rewrite the given problem using the appropriate operation symbol and solve:
   a. Find the difference of 20 and 5.
   b. Find the product of 2 and 4.

3. Is 7 a factor of 17?

4. Is 9 a prime or composite number?

5. Write the given factors in exponential notation:
   a. $3 \cdot 3 \cdot 3 \cdot 3 \cdot 2$
   b. Simplify: $4^2 - 2^3$

6. Simplify using order of operations:
   a. $20 \div 2 - 10 + 8 \, (2^2)$
   b. $(2 + 12) \div 7 + 3 \, (4) + 3^2$

7. Determine whether the following statement is true or false. If true, what property is being applied?
   a. $6 \div 3 = 3 \div 6$
   b. $(2 \cdot 3) \cdot 4 = 4 \cdot (3 \cdot 2)$

8. Are the following rational or irrational numbers:
   a. 7
   b. $\sqrt{7}$

9. If you buy a car for $5,820 and sell it the next year for $4,685, how much money did you lose?

CHECK YOUR ✓ ANSWERS — If you missed more than 3 answers, ask your instructor for help as soon as possible. Otherwise, you should be ready to take the topic quiz on Symbols and Definitions.

CHAPTER 3
PRIME NUMBERS

INTRODUCTION

In this chapter, you will learn the techniques for finding lowest common multiples (LCM) which in the next chapter—Chapter 4—will "suddenly" be called lowest common denominators (LCD).

SECTION 3-1 LISTING ALL OF THE FACTORS OF A NUMBER

Let's review what *factors* are. You should remember from Chapter 2 that factors are whole numbers which divide evenly (meaning no remainder) into other whole numbers.

EXAMPLE Factors of 20 are: 1, 2, 4, 5, 10, 20

All of the listed numbers divide evenly into 20. The list above also happens to be *all* of the factors of 20.

Another example:

Factors of 35 are: 1, 5, 7, 35

It's sometimes easier to think of factors in pairs. For instance, the first two factors of any number are the number itself and one. Take this last example—the factors of 35 are:

1 and 35 should be the first factors thought of since 1 times 35 equals 35. From there on it's a matter of trying to find other numbers (besides 1 and 35) which will divide evenly into 35.

Try dividing 2 into 35. Does it go evenly?

$$\begin{array}{r} 17 \\ 2\overline{)35} \\ \underline{2} \\ 15 \\ \underline{14} \\ 1 \end{array} \leftarrow \textbf{Remainder}$$

NO! 2 does not divide evenly into 35. Therefore, 2 is *not* a factor of 35. Try dividing 3 into 35. Does it go evenly?

$$\begin{array}{r} 11 \\ 3\overline{)35} \\ \underline{3} \\ 5 \\ \underline{3} \\ 2 \end{array} \leftarrow \textbf{Remainder}$$

NO! 3 does not divide evenly into 35. Therefore, 3 is *not* a factor of 35.
Try dividing 4 into 35.

$$\begin{array}{r} 8 \\ 4\overline{)35} \\ \underline{32} \\ 3 \end{array} \leftarrow \textbf{Remainder}$$

4 doesn't divide evenly into 35. So 4 is *not* a factor of 35.

Try dividing 5 into 35.

$$5\overline{)35} \begin{array}{r}7\\ \underline{35}\\ 0\end{array}$$

Ah! At last! 5 does divide evenly into 35. Therefore, 5 *and* the answer to our division (the quotient) 7 are both factors of 35. We have another pair of factors.

So far our list of factors of 35 is:

$$1, 35, 5, 7$$

The next number we should try to divide into 35 should be 6.

$$6\overline{)35} \begin{array}{r}5\\ \underline{30}\\ 5\end{array} \leftarrow \textbf{Remainder}$$

6 is *not* a factor of 35.

The next number we would try to divide into 35 is 7. However, since we already found that 7 is a factor of 35, we can stop at this point.

∴ Factors of 35 are: 1, 5, 7, 35.

When finding factors of numbers, start with the most obvious pair of factors: 1 and the number itself. Then start dividing by 2, 3, 4, and so on, listing as many pairs of factors as you can find until you reach a factor that you already have listed. Then you know it's time to stop.

GIVEN Find all the factors of 50.

Solution The first two factors to begin the list are:

$$1, 50$$

Now try dividing 2 into 50:

$$2\overline{)50} \begin{array}{r}25\\ \underline{4}\\ 10\\ \underline{10}\end{array}$$ Yes! 2 and 25 are factors of 50.

Our list is now:

$$1, 50, 2, 25$$

Try 3:

$$\begin{array}{r}16\\3{\overline{\smash{)}50}}\\\underline{3}\\20\\\underline{18}\\2\end{array}$$ ← **Remainder**

No! 3 is not a factor of 50.

Try 4:

$$\begin{array}{r}12\\4{\overline{\smash{)}50}}\\\underline{4}\\10\\\underline{8}\\2\end{array}$$ ← **Remainder**

No! 4 is not a factor of 50.

Try 5:

$$\begin{array}{r}10\\5{\overline{\smash{)}50}}\\\underline{5}\\0\\\underline{0}\end{array}$$

Yes! 5 and 10 are factors of 50.

Our list is now:

            1, 50, 2, 25, 5, 10

Trying 6, 7, 8, and 9 (a lot of the division can and should be done in your head—you don't have to write it down unless you feel it's necessary), you find that none of these are factors of 50. The next number to try is 10, which is already in our list of factors. Therefore, we know our list is complete. Now, rewrite the list so that the factors are in numerical order.

∴ All factors of 50 are: 1, 2, 5, 10, 25, 50

EXERCISE 3-1

Find all the factors of the following:

1. 2
2. 6
3. 9
4. 12
5. 15
*6. 18
7. 22
8. 28
9. 30
10. 45
*11. 60
12. 75
13. 90
*14. 92

SECTION 3-2 PRIME VERSUS COMPOSITE AND DIVISIBILITY TESTS

Do you recall the definition of a *prime number* from Chapter 2? A prime number has *only two* factors: itself and 1.

Examples of prime numbers, in numerical order, are:

$$2, 3, 5, 7, 11, 13, 17, 19 \ldots$$

(1 is *not* considered to be a prime number since it has only *one* factor—1.)

All of the numbers listed above have only themselves and 1 as factors. There are *no other numbers* which will divide evenly into these numbers.

Is 4 a prime number?
  No! 4 has *more* than 2 factors.
    ∴ 4 is a *composite* number. (See Chapter 2)

Is 9 a prime number?
  No! Factors of 9 are: 1, 3, 9
    ∴ 9 is a *composite* number.

Can you think of the next prime number after 19? Let's think. Is 20 prime? No! 20 has lots of factors besides itself and 1. Is 21 prime? No! 21 has factors of 3 and 7 besides 1 and 21. Is 22 prime? No! 22 has factors of 2 and 11 besides 1 and 21. Is 23 prime? Yes!! 23 only has factors of 1 and 23. So, *23 is the next prime after 19.*

Now our list of prime numbers is:

$$2, 3, 5, 7, 11, 13, 17, 19, 23 \ldots$$

Let's take a closer look at this list. Notice that the only even number on the list is our first number—2.

> **IMPORTANT:** 2 is the only *even* prime number. If you think about the rest of the even numbers such as: 4, 6, 8, 10, 12, 14, 16, 18, 20 and so on, they all can be evenly divided by 2, making 2 one of their factors. Therefore, no even number *with the exception of 2* can be prime since *2 will always be a factor of every even number.*

Let's take another look at our growing list of prime numbers. With the exception of 2, the rest of the prime numbers are odd numbers. However, *not all odd numbers are prime!*

Notice the list above does not contain 9, which is an odd number but *not* a prime number, since it has more than itself and 1 as factors (3 is also a factor of 9).

> **IMPORTANT:** Do *not* assume that just because a number is odd that it is a prime number. It may be prime, but then again, it may not be prime.

78 CHAPTER 3

How do you determine if a number is prime? By trying to prove that it isn't! In other words, try to find numbers (other than the given number and one) that will divide evenly into the number in question. If you find factors other than the given number and one, then the number is *not* prime. Remember, *a prime number can only be divided evenly by itself and one.* The numbers to use to try to divide into the number in question should be the list of prime numbers that you saw previously: 2, 3, 7, 11, 13 . . . Since we aren't trying to find a complete list of factors, just *a* number (other than the given number and one) that *will* divide evenly into the number in question, we only need to divide by the prime numbers.

GIVEN Is 51 a prime number?

Solution All we have to do is find a number that *will* divide evenly into 51. Once we find *one* number (besides 1 and 51) that works, we've proved that 51 is not a prime number. If we don't find any number that divides evenly into 51 other than 1 and 51, then we can say 51 is a prime.

The first number you should try to divide into 51 is the first prime number—2.

$$\begin{array}{r} 25 \\ 2\overline{)51} \\ \underline{4} \\ 11 \\ \underline{10} \\ 1 \end{array} \leftarrow \textbf{Remainder}$$

2 does *not* divide evenly into 51. Therefore, we must continue on. The next number to try is the next *prime* number—3.

$$\begin{array}{r} 17 \\ 3\overline{)51} \\ \underline{3} \\ 21 \\ \underline{21} \end{array}$$

Yes! 3 divides evenly into 51 and is therefore a factor of 51. This means that 51 is *not* a prime number. *51 is a composite number.*

GIVEN Is 29 a prime number?

Solution Start with 2:

$$\begin{array}{r} 14 \\ 2\overline{)29} \\ \underline{2} \\ 9 \\ \underline{8} \\ 1 \end{array} \leftarrow \textbf{Remainder}$$

Try 3:

$$\begin{array}{r} 9 \\ 3\overline{)29} \\ \underline{27} \\ 2 \end{array}$$ ← **Remainder**

Try 5, since it is the next prime on our list:

$$\begin{array}{r} 5 \\ 5\overline{)29} \\ \underline{25} \\ 4 \end{array}$$ ← **Remainder**

Try 7 (the next prime after 5):

$$\begin{array}{r} 4 \\ 7\overline{)29} \\ \underline{28} \\ 1 \end{array}$$ ← **Remainder**

Try 11 (the next prime after 7):

$$\begin{array}{r} 2 \\ 11\overline{)29} \\ \underline{22} \\ 7 \end{array}$$ ← **Remainder**

Try 13 (the next prime after 11):

$$\begin{array}{r} 2 \\ 13\overline{)29} \\ \underline{26} \\ 3 \end{array}$$ ← **Remainder**

By now you are probably getting tired of dividing into 29. It is a good assumption at this point to say that there are no other factors of 29 other than 29 and 1. Therefore, 29 is a prime number.

REMEMBER: If a number is not prime (if it has more factors besides itself and 1), then the number is called a composite number.

FOR THIS BOOK ONLY

IN GENERAL: To determine if a number is prime, try dividing the number by the primes from 2 through 13 (2, 3, 5, 7, 11, 13). If by the time you finish dividing by 13, you still haven't found a factor of the number other than itself and 1, assume for this book's purposes only, that the number is prime.

Wouldn't it be nice to have some tricks for telling just by looking at a number if it could be divided evenly by some other number? This would be especially useful when trying to determine if a large number is prime or not. Hooray!! We do have some tricks for telling if numbers can be evenly divided by some other numbers. These tricks are called *TESTS FOR DIVISIBILITY.*

We have tests for the first three prime numbers: 2, 3, and 5.

Divisibility Test for 2

STEP 1 Look at the last digit of the number in question (that is the right-most digit). If that last digit is an even number—0, 2, 4, 6, or 8—then the entire number is divisible by 2.

Examples:

68 is divisible by 2 since the last digit is an even number. (8 is the last digit and 8 is an even number.)

3,974 is divisible by 2 since its last digit is an even number. (4 is the last digit and 4 is an even number.)

9,562,770 is divisible by 2 since its last digit is an even number. (0 is the last digit and 0 is an even number.)

Divisibility Test for 3

STEP 1 Add the digits of the number together. If the sum can be evenly divided by 3, then the original number can be divided by 3.

Examples:

39 is divisible by 3. Adding the digits of 39 together, you get 3 + 9 = 12. Since 12 can be evenly divided by 3, so can 39 be evenly divided by 3.

78 is divisible by 3. Adding the digits of 78 together, you get 7 + 8 = 15. Since 15 can be evenly divided by 3, so can 78 be evenly divided by 3.

350 is *not* divisible by 3. Adding the digits of 350 together, you get 3 + 5 + 0 = 8. 8 cannot be evenly divided by 3. Therefore, 350 cannot be evenly divided by 3.

Divisibility Test for 5

STEP 1 If the last digit is 0 or 5, the entire number is divisible by 5.

Examples:

45 is divisible by 5. The last digit is a 5. Therefore, 45 can be divided evenly by 5.

905 is divisible by 5. The last digit in this number is a 5. Therefore, 905 can be divided evenly by 5.

2460 is divisible by 5. The last digit in this number is 0. Therefore, 2460 can be divided evenly by 5.

PRIME NUMBERS   81

Learn these three Tests for Divisibility. You will use them for many chapters in this book as well as for future mathematics.

Now let's look at some more numbers and try to determine if they are prime, using the Tests for Divisibility as much as possible.

GIVEN   Is 336 a prime number?

Solution   First try the test for 2:

Since 336 ends with an even digit (6 is an even number), 336 *can* be divided evenly by 2. Since 2 *is* a factor of 336, 336 cannot be prime.

∴ *336 is not prime.* 336 is a composite number.

GIVEN   Is 951 a prime number?

Solution   Try the test for 2 first:

951 does not end in an even digit (1 is *not* an even number). ∴ 2 is *not* a factor of 951. So far we have *not* proven that 951 is *not* prime.

Try the test for 3 next:

Adding the digits of 951 together, we get: 9 + 5 + 1 = 15. 15 *can* be divided evenly by 3. ∴ 951 can be divided by 3. That makes 3 a factor of 951 and ∴ *951 is not a prime number.* 951 is a composite number.

GIVEN   Is 97 a prime number?

Solution   Try the test for 2:

97 is *not* an even ending number. ∴ 2 is *not* a factor of 97.

Try the test for 3:

Adding the digits in 97 together, we get: 9 + 7 = 16. 16 *cannot* be evenly divided by 3. ∴ 3 is *not* a factor of 97.

Try the test for 5:

97 does not end in a 0 or 5. ∴ 5 is *not* a factor of 97.

So far we haven't been able to prove that 97 is *not* a prime number. We've also run out of tests for divisibility. Now we go back to long division.

Try 7:

97 is *not* evenly divided by 7:

$$\begin{array}{r} 13 \\ 7\overline{)97} \\ \underline{7} \\ 27 \\ \underline{21} \\ 6 \end{array}$$ ← **Remainder**   ∴ 7 is *not* a factor of 97.

Try 11:

97 is *not* evenly divided by 11:

$$\begin{array}{r} 8 \\ 11{\overline{\smash{\big)}\,97}} \\ \underline{88} \\ 9 \end{array} \leftarrow \text{Remainder}$$

∴ 11 is *not* a factor of 97.

Try 13:

13 is *not* a factor of 97:

$$\begin{array}{r} 7 \\ 13{\overline{\smash{\big)}\,97}} \\ \underline{91} \\ 6 \end{array} \leftarrow \text{Remainder}$$

We're up to 13 and still no factors for 97 have been found other than 1 and 97. At this point, we can say: *Yes! 97 is a prime number.*

EXERCISE 3-2

Determine whether the following numbers are prime or composite.

1. 27
2. 34
3. 57
4. 63
5. 75
6. 98
7. 105
8. 37
9. 59
10. 357
11. 121
12. 91
13. 143
14. 133
15. 159

PRIME NUMBERS   83

SECTION 3-3 PRIME FACTORING

Prime factorization of a number is the number expressed as a product of its prime factors. Prime factorization will eventually lead us into finding *Lowest Common Multiples,* which in turn helps us find *Lowest Common Denominators* in Chapter 4.

🖫 Procedure for Finding the Prime Factorization of a Number Using the Factor Tree

GIVEN   Find the prime factorization of 80.

Solution   **STEP 1**   Write the number and place an upside down V below it:

STEP 2   Using *Test for Divisibility first,* try to find a *prime* number that will divide evenly into the number in question. In this case, since the number ends in 0, we can either use a 2 or a 5 to divide evenly into 80. Let's use the 2 since it is the *first* prime we normally begin with. Write the 2 as shown below and place the quotient to the division of 80 by 2, also as shown.

STEP 3   Circle the 2, which says that 2 is prime. Since 40 is *not* prime, we do *not* circle it. Instead we place another upside down V below it.

STEP 4   Go back to Step 2, only this time try to find a prime number that will divide evenly into 40. 2 works again ∴ 40 ÷ 2 = 20.

Place the 2 and the 20 as shown on the right. Since 2 is prime, circle it. Since 20 is *not* prime, place an upside down V below it.

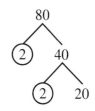

STEP 5 Go back to Step 2 again and try to find a prime that will divide evenly into 20. 2 works again. ∴ 20 ÷ 2 = 10. Place the new 2 and 10 as shown. Circle the 2 to show it's prime and since 10 is not prime, place an upside down V below it.

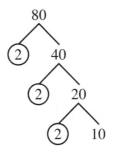

STEP 6 Go back to Step 2 and try to find a prime that divides evenly into 10. Since 10 ÷ 2 = 5, place the new 2 and 5 as shown.

Since 2 and 5 are prime, circle them both.

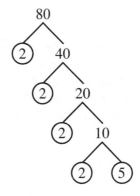

The prime factorization is complete since we have no more composite numbers to try to "break down" into prime factors.

STEP 7 State the prime factorization of the number as a product of the circled prime numbers. Then check your answer by multiplying the primes back together and see if you get the original number.

∴ Prime factorization of 80 = 2 • 2 • 2 • 2 • 5

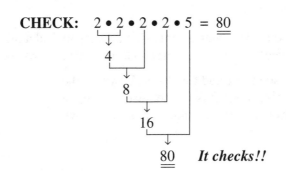

It checks!!

Let's try the *Factor Tree Method* again.

GIVEN   Find the prime factorization of 630.

Solution

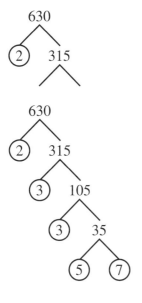

STEPS 1 & 2   Will 2 divide evenly into 630? Yes! 630 ÷ 2 = 315. Write 2 and 315 down and circle any prime number. Place the ∧ below any composite number.

STEP 2 AGAIN   Will 2 divide evenly into 315? No! Try 3. Yes! 315 ÷ 3 = 105.

STEP 2 AGAIN   Will 2 divide into 105 evenly? No! Try 3. Yes! 105 ÷ 3 = 35.

STEP 2 AGAIN   Will 2 divide into 35 evenly? No! Will 3? No! Try 5. Yes! 35 ÷ 5 = 7 Since both 5 and 7 (our last two factors) are prime, the prime factorization is complete.

∴ 630 = 2 • 3 • 3 • 5 • 7

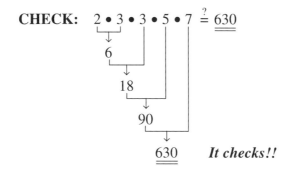

GIVEN   Find the prime factorization of 385.

Solution

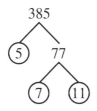

STEPS 1 & 2   Use Tests for Divisibility to see if 385 can be evenly divided by 2. No! By 3? No! By 5? Yes. 385 ÷ 5 = 77. 5 is prime—*circle it*. 77 is composite—*Break it down*. Use Tests for Divisibility to see what will divide evenly into 77. Try 2. No! Try 3. No! Try 5. No! Try the next prime—7. Yes. 77 ÷ 7 = 11. 7 is prime and so is 11 ∴ circle them both.

86 CHAPTER 3

$$\therefore 385 = \underline{\underline{5 \cdot 7 \cdot 11}}$$

CHECK: $5 \cdot 7 \cdot 11 \stackrel{?}{=} \underline{\underline{385}}$

$\quad\;\; 35$

$\quad\quad\quad\;\; \underline{\underline{385}}$ *Yes! It checks!!*

GIVEN Find the prime factorization of 2700.

Solution

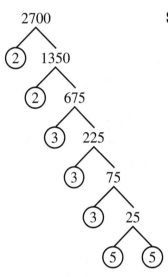

STEPS 1 & 2 Try 2. Yes! $2700 \div 2 = 1350$. Circle 2 because it is prime. Break down 1350 because it's composite.

Try 2. Yes! $1350 \div 2 = 675$. Circle 2. Break down 675.

Try 2. No! Try 3. Yes! $675 \div 3 = 225$. Circle 3. Break down 225.

Try 2. No! Try 3. Yes! $225 \div 3 = 75$. Circle 3. Break down 75.

Try 2. No! Try 3. Yes! $75 \div 3 = 25$. Circle 3. Break down 25.

Try 2. No! Try 3. No! Try 5. Yes. $25 \div 5 = 5$. Circle *both* 5's.

$$\therefore 2700 = \underline{\underline{2 \cdot 2 \cdot 3 \cdot 3 \cdot 3 \cdot 5 \cdot 5}}$$

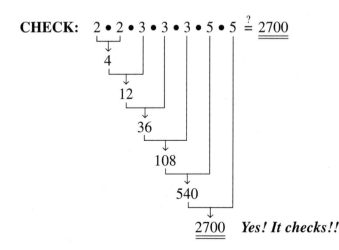

Yes! It checks!!

PRIME NUMBERS

REMEMBER: You only circle the *prime* factors, and make sure you *do* circle the *prime* factors. The prime factorization is the *product* of all the *circled* (prime) numbers in the factor tree.

Lest we forget—here is a list of the first ten prime numbers:

2
3
5
7
11
13
17
19
23
29

EXERCISE 3-3

Find the prime factorization of the following numbers, and check your answers.

1. 25
2. 18
3. 32
*4. 49
5. 63
6. 54
7. 72
*8. 95
9. 120
*10. 143
11. 200
*12. 350
13. 415
14. 568
15. 921

16. 715
17. 147
*18. 288
*19. 315
20. 1,000
21. 88
22. 16
23. 24
24. 38
25. 50
26. 69
27. 75
28. 36
29. 100
30. 81

SECTION 3-4 WHAT IS A LOWEST COMMON MULTIPLE?

Before this question can be answered, we have to first define what a *multiple* is. This is easily done through examples.

Muliples of 2 are:

$$2 \times 1 = 2$$
$$2 \times 2 = 4$$
$$2 \times 3 = 6$$
$$2 \times 4 = 8$$
$$2 \times 5 = 10$$
$$2 \times 6 = 12$$
$$2 \times 7 = 14$$

and so on

Multiples of a number are the number times the set of integers ("counting" numbers such as 1, 2, 3, 4, 5 . . .).

Multiples of 10 are:

$$10 \times 1 = 10$$
$$10 \times 2 = 20$$
$$10 \times 3 = 30$$
$$10 \times 4 = 40$$
$$\vdots \quad \vdots$$
$$10 \times 10 = 100$$
$$10 \times 11 = 110$$
$$\vdots \quad \vdots$$

and so on

Multiples of 25 are (as a list):   25, 50, 75, 100, 125, 150, 175 . . .

GIVEN   List the first 10 multiples of 3.

Solution

$$3 \times 1 = 3$$
$$3 \times 2 = 6$$
$$3 \times 3 = 9$$
$$3 \times 4 = 12$$
$$3 \times 5 = 15$$
$$3 \times 6 = 18$$
$$3 \times 7 = 21$$
$$3 \times 8 = 24$$
$$3 \times 9 = 27$$
$$3 \times 10 = 30$$

∴ The first ten multiples of 3 are: 3, 6, 9, 12, 15, 18, 21, 24, 27, 30

PRIME NUMBERS

GIVEN   List the first 5 multiples of 9

Solution
$9 \times 1 = \mathbf{9}$
$9 \times 2 = \mathbf{18}$
$9 \times 3 = \mathbf{27}$
$9 \times 4 = \mathbf{36}$
$9 \times 5 = \mathbf{45}$

∴ The first 5 multiples of 9 are: 9, 18, 27, 36, 45

GIVEN   List the first 3 multiples of 200.

Solution
$200 \times 1 = \mathbf{200}$
$200 \times 2 = \mathbf{400}$
$200 \times 3 = \mathbf{600}$

∴ The first 3 multiples of 200 are: 200, 400, 600

Make sure you see the difference between *factors* and multiples:

Factors of 10 are 1, 2, 5, 10

Multiples of 10 are 10, 20, 30, 40, 50 . . .

Factors of a number are *never* larger than the number itself. The largest *factor* of 10 (as seen above) *is* 10.

Multiples of a number are *never* smaller than the number itself. The smallest multiple of 10 *is* 10. The largest multiple of 10 is . . . Think about it!

EXERCISE 3-4

List the first 5 multiples of the following.

*1. 2
2. 3
3. 4
4. 5
5. 6
*6. 7
7. 8
8. 9
9. 10
10. 11

11. 12
12. 13
13. 14
14. 15
15. 16
*16. 17
17. 18
18. 19
19. 20
20. 50

SECTION 3-5 FINDING THE LOWEST COMMON MULTIPLE

Are we ready to talk about *Lowest Common Multiple* (LCM)? Not quite. We first need to talk about common multiples. If you list the first 10 multiples of 2 *and* the first 10 multiples of 3, you'll find that 2 and 3 *share* some of the same multiples. In other words, they have some *common multiples*.

First 10 multiples of 2 are:

2, 4, **6,** 8, 10, **12,** 14, 16, **18,** 20

First 10 multiples of 3 are:

3, **6,** 9, **12,** 15, **18,** 21, 24, 27, 30

Both 2 and 3 have the multiples 6, 12, and 18 in common, just in listing their first 10 multiples. (Imagine how many multiples they have in common after listing the first 100 multiples of both.)

Notice that the *first* common multiple for 2 and 3 is 6. 6 is the *Lowest Common Multiple*—6 is the *smallest* multiple that 2 and 3 share.

> *Lowest Common Multiple* or *LCM* is the smallest **multiple** that two or more numbers have in common.

Let's take a look at some more LCM's.

GIVEN Find the LCM of 4 and 10.

Solution Using the method we used before for finding common multiples of 2 and 3, we list the multiples of 4 and 10, and compare.

Multiples of 4 = 4, 8, 12, 16, **20,** 24, 28, 32, 36 . . .

Multiples of 10 = 10, **20,** 30, 40, 50, 60, 70, 80 . . .

20 is the *first common* multiple we find in both lists ∴ 20 is the Least Common Multiple for 4 and 10.

GIVEN Find the LCM for 8, 15 and 21.

Solution Here is where the previous method for finding LCM falls flat. It may take us a very long time to list as many multiples as we would need in order to find a common multiple for 8, 15, and 21. Is there an easier way for finding LCM's that works *everytime*, no matter how many numbers or how big the numbers are? *Yes!!*

Procedure for Finding LCM of Two or More Numbers

STEP 1 Find the prime factorization of each number.

STEP 2 Construct a box with a line for each number; then make smaller boxes—as many as you think you might need, you can always add more.

STEP 3 Starting with the first number, place *all* of its prime factors across from the first number.

STEP 4 Going on to the second number, place all of its prime factors across from it, *but* try to line up any like prime factors that you listed from the first number.

STEP 5 Repeat Step 4 for the third number and the fourth number and so on.

STEP 6 Bring down one prime factor from each column in the larger box.

STEP 7 Multiply the prime factors you brought down in Step 6 together. The product is the *LCM*.

Now it's time to make some sense out of these steps.

GIVEN Find the LCM of 8, 15, and 21.

Solution **STEP 1** Prime factor 8, 15, and 21

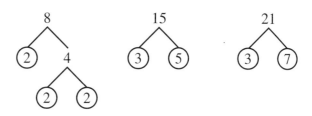

STEP 2 Construct box with 3 lines since we have 3 numbers we're working with— 8, 15, and 21.

This line will be for 8

This line will be for 15

This line will be for 21

Now make smaller boxes inside this large box by adding some vertical lines or columns.

8

15

21

Make sure you *always* add these vertical lines!

STEP 3 Place all of the prime factors of 8 across from 8 in the large box; each prime factor gets its own little box.

8	2	2	2			
15						
21						

STEP 4 Place all of the prime factors of 15 across from 15, but line up any like primes that are listed above.

8	2	2	2			
15				3	5	
21						

In this case, there were no like prime factors so 3 and 5 go to the first clear column across from 15.

STEP 5 Place all of the prime factors of 21 across from 21, but line up any like primes that have been used previously.

8	2	2	2			
15				3	5	
21				3		7

This time the 3 was able to line up with the 3 we listed as one of 15's prime factors. However, we couldn't line up the 7 so it moved over to the first clear column.

STEP 6 Bring down *one* prime factor from each column.

8	2	2	2			
15				3	5	
21				3		7

↓ ↓ ↓ ↓ ↓ ↓
2 • 2 • 2 • 3 • 5 • 7

PRIME NUMBERS 93

STEP 7   Multiply the prime factors brought down in Step 6 together. This is the LCM.

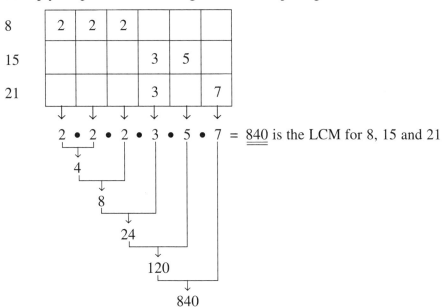

∴ 840 is the smallest *multiple* that 8, 15, and 21 have in common.

∴ LCM for 8, 15, and 21 = 840

GIVEN   Find LCM for 20, 45, 48, and 75

Solution   **STEP 1**   Prime factor 20, 45, 48, and 75

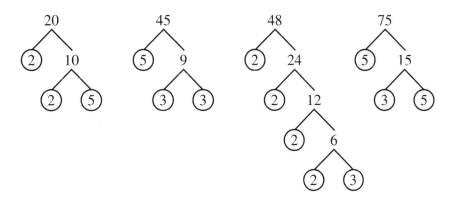

STEP 2   Construct box with 4 lines since we have 4 numbers we're working with—20, 45, 48, and 75. Then make smaller boxes by adding vertical lines.

This line is for 20

This line is for 45

This line is for 48

This line is for 75

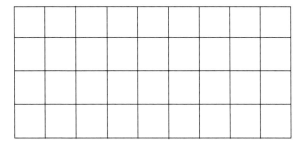

STEP 3 Place all of the prime factors of 20 across from 20 in the large box; each prime factor of 20 getting its own little box.

20	2	2	5				

STEP 4 Place all of the prime factors of 45 across from 45, but line up any like primes that are listed above (across from 20).

20	2	2	5				
45			5	3	3		

In this case, the 5 was able to be lined up with the 5 listed above as one of 20's prime factors. There were no 3's used above ∴ the 3's were placed in first clear columns across from 45.

STEP 5 Place all of the prime factors of 48 across from 48, and line up any like primes used above.

20	2	2	5				
45			5	3	3		
48	2	2		3		2	2

Two of the 2's were able to be lined up, as was the 3. The other two 2's were placed in first clear columns across from 48.

Repeat Step 5 for 75.

20	2	2	5				
45			5	3	3		
48	2	2		3		2	2
75			5	3			5

One of the 5's was able to be lined up, as was the 3. The second 5 was placed in the first clear column across from 75.

STEP 6   Bring down *one* prime factor from each column.

	2	2	5					
20	2	2	5					
45			5	3	3			
48	2	2		3		2	2	
75			5	3				5

↓ ↓ ↓ ↓ ↓ ↓ ↓ ↓
2 • 2 • 5 • 3 • 3 • 2 • 2 • 5

STEP 7   Multiply the prime factors brought down in Step 6 together. This product is the LCM.

2 • 2 • 5 • 3 • 3 • 2 • 2 • 5 = 3,600 is the LCM for 20, 45, 48 and 75

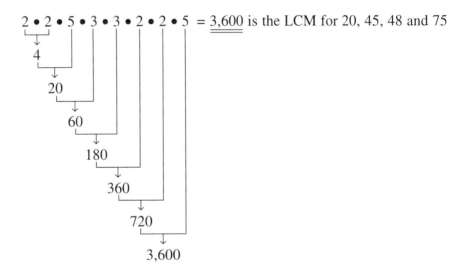

∴ 3,600 is the smallest *multiple* that 20, 45, 48, and 75 have in common.

GIVEN   Find LCM for 20, 32, and 45.

Solution   **STEP 1**

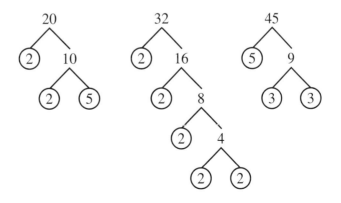

96 CHAPTER 3

STEPS 2, 3, 4, 5, 6, AND 7

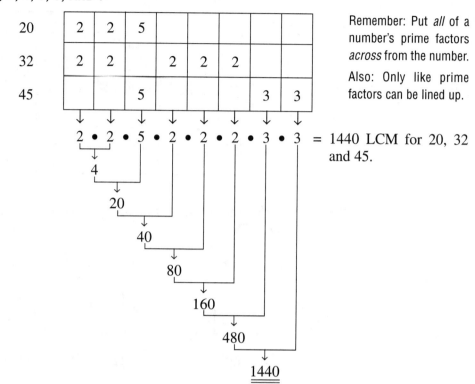

∴ 1440 is the smallest multiple that 20, 32, and 45 have in common.

LCM for 20, 32, and 45 = 1440

GIVEN Find the LCM for 4, 6, and 24.

Solution **STEP 1**

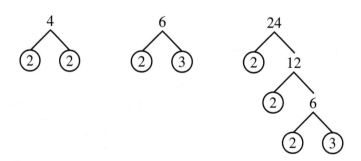

STEPS 2, 3, 4, 5, 6, AND 7

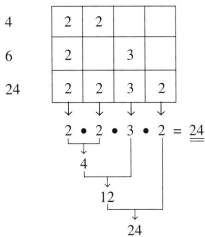

Notice that if lining up like prime factors means changing the order of a number's prime factors, that's *OK*, because of the Commutative Property of Multiplication.

∴ $\underline{\underline{24}}$ is the LCM for 4, 6, and 24.

GIVEN   Find the LCM of 39 and 65.

Solution   **STEP 1**

STEPS 2, 3, 4, 6, AND 7

39	3	13	
65		13	5

3 • 13 • 5 = $\underline{\underline{195}}$ is the LCM for 39 and 65.

   39
   195

∴ LCM for 39 and 65 = $\underline{\underline{195}}$.

GIVEN   Find LCM for 9, 12, 28, and 35.

Solution   **STEP 1**

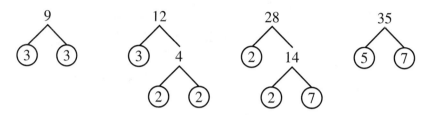

STEPS 2, 3, 4, 5, 6 AND 7

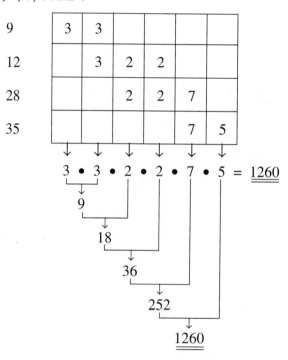

∴ LCM for 9, 12, 28, and 35 = 1260.

GIVEN   Find the LCM of 5, 7, and 13.

Solution   **STEP 1**

⑤   ⑦   ⑬

STEPS 2, 3, 4, 6, AND 7

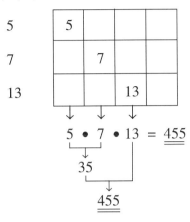

∴ LCM of 5, 7, and 13 = 455.

EXERCISE 3-5

Find the LCM's of the given numbers.

*1. 4 and 10
2. 9 and 18
3. 25 and 30
4. 12 and 16
5. 22 and 33
*6. 17 and 34
7. 13 and 52
8. 5 and 11
9. 6 and 9
*10. 8, 12, and 15
11. 9, 21 and 30
*12. 5, 15, and 30
13. 4, 14, and 35
*14. 16, 18 and 20
15. 16, 32, and 64

16. 4, 8, and 24
*17. 17, 34, and 40
18. 13, 15, and 39
19. 20, 30, and 40
20. 15, 25, and 50
*21. 4, 6, 12, and 15
22. 8, 9, 18, and 24
23. 7, 9, 11, and 13
24. 2, 4, 8, and 16
25. 3, 6, 9, and 12
26. 5, 10, 15, and 30
*27. 25, 50, 75, and 100
28. 30, 35, 40, and 45
29. 14, 21, 35, and 42
*30. 10, 100, 1,000, and 10,000

IMPORTANT: The method you have just learned for finding LCM's is the same method you will use in the next chapter for finding lowest common denominators (LCD). So make sure you understand this procedure *before* you continue on.

CHAPTER 3

Review Test #1

1. List all the factors of: (see 3-1)
    a. 16
    b. 72

2. Are any of the numbers from 35 to 45 prime? If yes, what are they? (see 3-2)

3. Find the prime factorization of: (see 3-3)
    a. 28
    b. 125
    c. 2010

4. Find the LCM for: (see 3-5)
    a. 15 and 25
    b. 8, 18, and 27
    c. 22, 49, and 105

 If you missed more than two (2) problems, correct your mistakes by working through this chapter again. Then take Review Test #2. If you missed two or fewer problems, correct your mistakes and ask your instructor for the topic quiz on Prime Numbers.

CHAPTER 3

Review Test #2

1. List all the factors of:
   a. 21
   b. 56

2. Are any of the numbers from 45 to 55 prime? If yes, what are they?

3. Find the prime factorization of:
   a. 36
   b. 105
   c. 356

4. Find the LCM for:
   a. 13 and 39
   b. 21, 25, and 30
   c. 17, 49, and 238

CHECK YOUR ✓ ANSWERS — If you missed more than two (2) problems, correct your mistakes and take Chapter Review Test #3. Otherwise, ask your instructor for the topic quiz on Prime Numbers.

CHAPTER 3

REVIEW TEST #3

1. List all the factors of:
   a. 36
   b. 110

2. Are any of the numbers from 25 to 35 prime? If yes, what are they?

3. Find the prime factorization for:
   a. 40
   b. 150
   c. 484

4. Find the LCM for:
   a. 16 and 40
   b. 9, 16, and 32
   c. 24, 30, and 32.

 CHECK YOUR ANSWERS If you missed more than two problems ask your instructor for help as soon as possible. Otherwise, you should be ready to ask your instructor for the quiz on Prime Numbers.

CHAPTERS 1-3

Cumulative Review #1

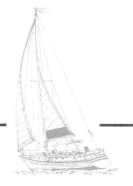

Use the **Cumulative Reviews** that are spaced throughout this book to help you review and retain the math skills that you have gained. For each of the eight chapters tested on the **Arithmetic Proficiency Test,** you will be required to correctly answer four out of five questions from each chapter. See how you would do with these first three chapters.

Whole Numbers

1. $5208 + 49 + 657 + 2$
2. $3439 - 148$
3. 295×463
4. $139363 \div 463$
5. $4292 \div 29$

Symbols and Definitions

1. Sum means to _____ ? (2-2)
2. Simplify: $8^2 + 2^3$ (2-4)
3. Write using exponential notation: $3 \cdot 3 \cdot 3 \cdot 4 \cdot 4 \cdot 7$ (2-4)
4. Simplify using order of operations: $3^2 \cdot (4 - 2) \div 6 + 2(18 + 3)$ (2-5)
5. Is $\frac{3}{4}$ a rational or irrational number? (2-7)

Prime Numbers

1. Write the prime factorization of: (3-3)
   a. 300 _____   b. 565 _____
2. Write all of the factors of 120 _____ (3-1)
3. Find the Least Common Multiple (LCM) of: (3-5)
   a. 16, 30, 36 _____   b. 20, 45, 150 _____

REMEMBER TO CHECK YOUR ANSWERS WITH APPENDIX A IN THE BACK OF THE BOOK.

CHAPTER 4

FRACTIONS I: ADDITION AND SUBTRACTION

INTRODUCTION

In this chapter you will become reacquainted with fractions: proper and improper fractions, mixed numbers, equivalent fractions, reduction of fractions, addition and subtraction of fractions, and also the concept of *rational numbers*.

SECTION 4-1 WHAT IS A FRACTION?

We are all familiar with the idea of a fraction representing equal parts of a whole. For example:

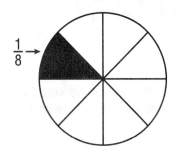

Divide a pie into eight equal parts, remove (eat) seven pieces and you have one piece remaining. Therefore, we have $\frac{1}{8}$ of the pie remaining. The denominator, or the bottom number, tells us how many equal pieces the whole pie was cut into. The numerator, or the top number, tells us how many pieces we have out of the total.

This concept of fractions is the one we all grew up with; but now let's look at fractions in a slightly different light: as *rational numbers*.

Rational numbers are numbers that can be expressed as the division of one whole number by another whole number. For instance, if we are asked to divide 3 by 4, we could show the division in these ways:

$3 \div 4$ read—3 divided by 4

$4\overline{)3}$ read—4 divided into 3

$\frac{3}{4}$ read—three fourths

All three of these ways of expressing the division of whole number 3 by whole number 4 are correct. The fraction, $\frac{3}{4}$, is a *rational number*. It is the *ratio* (or comparison by division) of whole number, 3, to whole number, 4. So $\frac{3}{4}$ is another way of saying 3 divided by 4.

Some additional examples of expressing whole number division as fractions are:

$5 \div 7 = \frac{5}{7}$ read—five seven*ths*

$6 \div 20 = \frac{6}{20}$ read—six twentie*ths*

$35 \div 100 = \frac{35}{100}$ read—thirty-five hundred*ths*

$18 \div 7 = \frac{18}{7}$ read—eighteen seven*ths*

(Notice the *denominator* of a fraction is read with a "*ths*" ending.)

This idea of looking at fractions as rational numbers, or as the comparison of two whole numbers by division will come in very handy when you begin the chapters on decimals, and later when you begin to "explore" algebra, and later still if you continue into trigonometry and calculus. In other words, fractions are here to stay. So let's learn how to work with them *now*.

SECTION 4-2 FRACTION TERMS

Let's review some terms used in working with fractions:

Term	Definition	Example
Numerator	The number on top of the fraction line tells us how many pieces of the whole that we have	$\frac{5}{7}$, 5 is the numerator—we have 5 of 7 pieces $\frac{8}{3}$, 8 is the numerator.
Denominator	The number on the bottom of the fraction, also known as the divisor—tells us how many pieces the whole was originally cut into	$\frac{6}{9}$, 9 is the denominator—the whole was cut into 9 pieces $\frac{20}{3}$, 3 is the denominator (divisor).
Proper Fraction	Fraction in which the numerator is *smaller than* the denominator (numerator < denominator)	$\frac{2}{3}$ is a proper fraction since 2 < 3. $\frac{10}{13}$ is a proper fraction since 10 < 13.
Improper Fraction	Fraction in which the numerator is larger than or equal to the denominator (numerator > denominator)	$\frac{8}{5}$ is an improper fraction since 8 > 5. $\frac{100}{9}$ is an improper fraction since 100 > 9. $\frac{6}{6}$ is an improper fraction since 6 = 6.

EXERCISE 4-2

In problems 1–20 identify the numerator, denominator, and whether the fraction is proper or improper. Example: $\frac{5}{3}$, 5 is the numerator, 3 is the denominator; the fraction is improper since 5 > 3.

1. $\frac{4}{5}$
2. $\frac{10}{13}$
3. $\frac{5}{8}$
4. $\frac{9}{8}$
5. $\frac{26}{27}$
6. $\frac{18}{15}$
7. $\frac{200}{20}$
8. $\frac{390}{400}$
9. $\frac{1}{2}$
10. $\frac{19990}{19999}$
11. $\frac{8}{13}$
12. $\frac{29}{28}$
13. $\frac{15}{15}$
14. $\frac{42}{1}$
15. $\frac{87}{87}$
16. $\frac{3990}{3990}$
17. $\frac{1}{856}$
18. $\frac{856}{1}$
19. $\frac{3}{3}$
20. $\frac{1,000,000}{1,000,000}$

SECTION 4-3 CHANGING IMPROPER FRACTIONS INTO MIXED NUMBERS AND BACK AGAIN

Term	Definition	Example
Mixed Number	A mixed number represents the sum of a whole number and a fraction.	2 plus $\frac{1}{2}$ = $2\frac{1}{2}$ (read—two *and* one half) 13 plus $\frac{4}{9}$ = $13\frac{4}{9}$ (read—thirteen *and* four nineths) 100 plus $\frac{5}{16}$ = $100\frac{5}{16}$ (read—one hundred *and* five sixteenths)

> All improper fractions can be expressed as mixed numbers or whole numbers and all mixed numbers and whole numbers can be expressed as improper fractions.

Procedure for Changing an Improper Fraction into a Mixed Number

GIVEN Write $\frac{7}{4}$ as a mixed number.

Solution

STEP 1
Divide the numerator by the denominator.

$$\begin{array}{r} 1 \\ 4\overline{)7} \\ -4 \\ \hline 3 \end{array}$$ ← whole number part of answer

← remainder

STEP 2
Write the remainder over the divisor.

$$1\frac{3}{4}$$ ← remainder
← divisor

$$\begin{array}{r} 4\overline{)7} \\ -4 \\ \hline 3 \end{array}$$

STEP 3
Write the answer.

$$\therefore \frac{7}{4} = 1\frac{3}{4}$$

GIVEN Write $\frac{25}{6}$ as a mixed number.

Solution

STEP 1

$$\begin{array}{r} 4 \\ 6\overline{)25} \\ 24 \\ \hline 1 \end{array}$$

STEP 2

$$4\frac{1}{6}$$

$$\begin{array}{r} 6\overline{)25} \\ 24 \\ \hline 1 \end{array}$$

STEP 3

$$\therefore \frac{25}{6} = 4\frac{1}{6}$$

FRACTIONS I: ADDITION AND SUBTRACTION   **111**

GIVEN   Write $\dfrac{14}{3}$ as a mixed number.

Solution   **STEPS 1 AND 2**

$$\begin{array}{r} 4\frac{2}{3} \\ 3\overline{)14} \\ \underline{12} \\ 2 \end{array}$$

STEP 3

$$\therefore \dfrac{14}{3} = 4\dfrac{2}{3}$$

▣ Procedure for Changing Mixed Numbers into Improper Fractions

GIVEN   Write $9\dfrac{5}{8}$ as an improper fraction.

Solution

STEP 1

Multiply the denominator of the fractional part by the whole number part.

$$9\dfrac{5}{8}$$

$(8 \times 9 = 72)$

STEP 2

Add the numerator of the fractional part to the product from Step 1.

$$9\dfrac{5}{8}$$

$[(72) + 5 = 77]$

STEP 3

The sum from Step 2 is placed over the denominator.

$$9\dfrac{5}{8} = \dfrac{77}{8}$$

GIVEN   Write $4\dfrac{3}{5}$ as an improper fraction.

Solution   **STEPS 1, 2, AND 3**

$$4\dfrac{3}{5} = \dfrac{(5 \times 4) + 3}{5} = \dfrac{23}{5}$$

GIVEN   Write $10\dfrac{1}{6}$ as an improper fraction.

Solution   **STEPS 1, 2, AND 3**

$$10\dfrac{1}{6} = \dfrac{(6 \times 10) + 1}{6} = \dfrac{61}{6}$$

112 CHAPTER 4

GIVEN Write 5 as an improper fraction.

Solution Since 5 has no fraction with it we simply put 5 over 1: $\frac{5}{1}$

$\therefore 5 = \frac{5}{1}$

All whole numbers can be made to look like fractions by placing the whole number over 1.

EXAMPLES $6 = \frac{6}{1}; \quad 48 = \frac{48}{1}; \quad 100 = \frac{100}{1}$

EXERCISE 4-3

In problems 1–20, change the improper fractions into mixed numbers.

1. $\frac{5}{3}$
2. $\frac{9}{5}$
3. $\frac{21}{8}$
4. $\frac{32}{7}$
5. $\frac{60}{9}$
6. $\frac{75}{4}$
7. $\frac{128}{3}$
8. $\frac{500}{11}$
9. $\frac{4052}{13}$
10. $\frac{131}{2}$
11. $\frac{51}{3}$
12. $\frac{16}{1}$
13. $\frac{18}{9}$
14. $\frac{17}{8}$
15. $\frac{9}{9}$
16. $\frac{72}{8}$
17. $\frac{150}{1}$
18. $\frac{65}{5}$
19. $\frac{18}{18}$
20. $\frac{100}{100}$

In problems 21–40, change the mixed numbers or whole numbers into improper fractions.

21. $4\frac{2}{3}$

22. $8\frac{1}{6}$

23. $2\frac{1}{3}$

24. $6\frac{5}{8}$

25. $7\frac{2}{9}$

26. $14\frac{1}{2}$

27. $16\frac{5}{6}$

28. $20\frac{9}{13}$

29. $43\frac{3}{4}$

30. $161\frac{1}{8}$

31. $5\frac{16}{21}$

32. $24\frac{31}{29}$

33. $10\frac{55}{62}$

34. $22\frac{2}{22}$

35. $81\frac{1}{18}$

36. 6

37. 15

38. 27

39. 101

40. 1

114 CHAPTER 4

4-4 REDUCING FRACTIONS

Fraction answers must always be in the lowest terms or *REDUCED*.
Let's find out the procedure to get equivalent fractions, but expressed in lower terms—known as *reducing fractions to simplest form.*

A Method for Reducing Fractions

STEP 1 Find a *common factor* for the numerator and denominator of the fraction to be reduced, using *tests for divisibility* (3-2).

STEP 2 Divide the numerator and denominator by their common factor.

STEP 3 Check the reduced fraction to see if there are any more common factors for its numerator and denominator. If there are, then go back to Step 2. If there are no more common factors, then you have the equivalent fraction in simplest form.

GIVEN Reduce $\frac{25}{30}$ to simplest form.

Solution $\frac{25}{30} = \frac{?}{?}$

STEP 1: By using tests for divisibility, we can see that 5 is a common factor of 25 and 30.

$\frac{25 \div 5}{30 \div 5} = \frac{?}{?}$

STEP 2: Divide the numerator and denominator by 5.

$\frac{25}{30} \div \left(\frac{5}{5}\right) = \frac{5}{\underline{\underline{6}}}$

STEP 3: Check reduced fraction ($\frac{5}{6}$) for additional common factors. There are none.
$\therefore \frac{25}{30} = \frac{5}{6}$

GIVEN Reduce $\frac{18}{20}$ to simplest form.

Solution $\frac{18}{20} = \frac{?}{?}$

STEP 1: 2 is a common factor of 18 and 20.

$\frac{18 \div 2}{20 \div 2} = \frac{?}{?}$

STEP 2: Divide 18 and 20 by 2.

$\frac{18}{20} \div \left(\frac{2}{2}\right) = \frac{9}{\underline{\underline{10}}}$

STEP 3: There are no more common factors shared by 9 and 10.
$\therefore \frac{18}{20} = \frac{9}{10}$

FRACTIONS I: ADDITION AND SUBTRACTION

Notice that we are dividing our original fraction by a fraction that is really 1 ($\frac{2}{2} = 1$). Dividing by 1 does not change the value of the fraction just as multiplying by 1 does not change the value; only the way the fraction "looks" is changed.

GIVEN Reduce $\frac{18}{24}$ to simplest form.

Solution

$\frac{18}{24} = \frac{?}{?}$

STEP 1: 2 is a common factor of 18 and 24.

$\frac{18 \div 2}{24 \div 2} = \frac{9}{12}$

STEP 2: Divide 18 and 24 by 2.

$\frac{18}{24} = \frac{9}{12} = \frac{?}{?}$

STEP 3: 3 is a common factor of 9 and 12.

$\frac{18}{24} = \frac{9 \div 3}{12 \div 3} = \frac{3}{4}$

Back to Step 2—divide 9 and 12 by 3.

$\therefore \frac{18}{24} = \frac{3}{4}$

STEP 3: There are no more common factors shared by 3 and 4.
$\therefore \frac{18}{24} = \frac{3}{4}$

GIVEN Reduce $\frac{60}{44}$ to simplest form.

Solution

$\frac{60}{44} = \frac{?}{?}$

STEP 1: 2 is a common factor of 60 and 44.

$\frac{60 \div 2}{44 \div 2} = \frac{30}{22}$

STEP 2: Divide 60 and 44 by 2.

$\frac{60}{44} = \frac{30}{22} = \frac{?}{?}$

STEP 3: 2 is a common factor of 30 and 22.
$\therefore$ back to Step 2.

$\frac{60}{44} = \frac{30 \div 2}{22 \div 2} = \frac{15}{11}$

STEP 2: Divide 30 and 22 by 2.

$\therefore \frac{60}{44} = \frac{15}{11}$

STEP 3: There are no more common factors shared by 15 and 11.

🗐 An Alternate Method for Reducing Fractions

STEP 1   Do a prime factorization of the numerator and denominator.

STEP 2   Using Commutative Property of Multiplication, change the order of the prime factors in the denominator such that like primes in the numerator are directly over like primes in the denominator.

STEP 3   Cancel the like primes.

STEP 4   Multiply the primes left in the numerator together and then multiply the primes left in the denominator together.

GIVEN   Reduce $\dfrac{60}{126}$ to simplest form.

Solution

$\dfrac{60}{126} = \dfrac{2 \cdot 2 \cdot 3 \cdot 5}{3 \cdot 3 \cdot 2 \cdot 7}$ **STEP 1:** Do prime factorization of numerator and denominator.

$\dfrac{60}{126} = \dfrac{2 \cdot 2 \cdot 3 \cdot 5}{2 \;\;\cdot 3 \;\;\;\cdot 3 \cdot 7}$ **STEP 2:** Line up like primes.

$\dfrac{60}{126} = \dfrac{\cancel{2} \cdot 2 \cdot \cancel{3} \cdot 5}{\cancel{2} \;\;\cdot \cancel{3} \;\;\;\cdot 3 \cdot 7}$ **STEP 3:** Cancel the like primes.

$\dfrac{60}{126} = \dfrac{1 \cdot 2 \cdot 1 \cdot 5}{1 \;\;\cdot 1 \;\;\;\cdot 3 \cdot 7} = \dfrac{10}{21}$ **STEP 4:** Multiply primes left in numerator together and primes left in the denominator together (although 1 is not a prime number, when we cancel out like primes they must be replaced with 1.)

$\therefore \dfrac{60}{126} = \underline{\underline{\dfrac{10}{21}}}$

NOTE   **Always assume that a fraction *answer* must be in *reduced form*.**

FRACTIONS I: ADDITION AND SUBTRACTION **117**

GIVEN Reduce $\dfrac{5}{45}$ to simplest form.

Solution

$\dfrac{5}{45} = \dfrac{5}{3 \cdot 3 \cdot 5}$ **STEP 1:** Prime factor numerator and denominator.

$\dfrac{5}{45} = \dfrac{5}{5 \cdot 3 \cdot 3}$ **STEP 2:** Line up like primes.

$\dfrac{5}{45} = \dfrac{\cancel{5}}{\cancel{5} \cdot 3 \cdot 3}$ **STEP 3:** Cancel like primes.

$\dfrac{5}{45} = \dfrac{1}{1 \cdot 3 \cdot 3} = \dfrac{1}{9}$ **STEP 4:** Multiply factors left in numerator together and multiply factors left in denominator together.

$\dfrac{5}{45} = \underline{\underline{\dfrac{1}{9}}}$

GIVEN Reduce $\dfrac{60}{4}$ to simplest form.

Solution

$\dfrac{60}{4} = \dfrac{2 \cdot 3 \cdot 2 \cdot 5}{2 \cdot 2}$ **STEP 1:** Prime factor numerator and denominator.

$\dfrac{60}{4} = \dfrac{2 \cdot 3 \cdot 2 \cdot 5}{2 \quad \cdot 2}$ **STEP 2:** Line up like primes.

$\dfrac{60}{4} = \dfrac{\cancel{2} \cdot 3 \cdot \cancel{2} \cdot 5}{\cancel{2} \quad \cdot \cancel{2}}$ **STEP 3:** Cancel.

$\dfrac{60}{4} = \dfrac{1 \cdot 3 \cdot 1 \cdot 5}{1 \quad \cdot 1} = \dfrac{15}{1}$ **STEP 4:** Multiply factors in numerator together, then multiply factors in denominator together. Since $\tfrac{15}{1}$ *is* usually written as a whole number:

$\dfrac{60}{4} = \underline{\underline{\dfrac{15}{1} = 15}}$

GIVEN Reduce $\dfrac{0}{18}$ to simplest form.

Solution

$\dfrac{0}{18} = \underline{\underline{0}}$ **REMEMBER:** A fraction is another way of expressing division; in this case, division of 0 by 18. Therefore, reducing this fraction is to answer the problem:

$$0 \div 18 = ? \quad \therefore \quad \dfrac{0}{18} = 0$$

EXERCISE 4-4

Reduce all the following to simplest form using either method.

*1. $\dfrac{12}{15}$

2. $\dfrac{9}{24}$

3. $\dfrac{15}{30}$

4. $\dfrac{21}{28}$

*5. $\dfrac{25}{75}$

6. $\dfrac{33}{90}$

7. $\dfrac{18}{20}$

8. $\dfrac{32}{64}$

*9. $\dfrac{90}{180}$

*10. $\dfrac{25}{20}$

11. $\dfrac{260}{40}$

12. $\dfrac{15}{3}$

13. $\dfrac{200}{36}$

*14. $\dfrac{1875}{4}$

15. $\dfrac{29}{58}$

*16. $\dfrac{7}{105}$

17. $\dfrac{48}{35}$

18. $\dfrac{66}{220}$

19. $\dfrac{192}{640}$

*20. $\dfrac{81}{168}$

21. $\dfrac{144}{36}$

22. $\dfrac{9}{90}$

23. $\dfrac{48}{144}$

*24. $\dfrac{0}{44}$

25. $\dfrac{55}{77}$

26. $\dfrac{16}{12}$

*27. $\dfrac{75}{25}$

28. $\dfrac{180}{90}$

*29. $\dfrac{0}{15}$

30. $\dfrac{80}{45}$

SECTION 4-5 BUILDING EQUIVALENT FRACTIONS

We have learned how to reduce fractions. Now we must learn how to build equivalent fractions. This is the reverse of reducing and must be done whenever we add or subtract fractions that do *not* have *like* denominators. Follow these examples very carefully. This will become the second step in adding or subtracting fractions with *un*like denominators.

Term	Definition	Example
Equivalent Fractions	Fractions that are equal in value but have different denominators (and numerators).	$\frac{1}{2} = \frac{2}{4}$
		$\frac{3}{4} = \frac{6}{8}$
		$\frac{2}{3} = \frac{4}{6}$
		$\frac{8}{10} = \frac{4}{5}$

Let's look at the first example, $\frac{1}{2} = \frac{2}{4}$, and go back for a moment to check our old method of looking at fractions: as equal parts of a whole. Divide a line into two equal parts.

Divide a second line, *of equal length with the first,* into four equal parts.

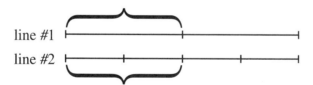

$\frac{1}{2}$ of the first line covers the same distance as $\frac{2}{4}$ of the second line; ∴ $\frac{1}{2}$ of line #1 = $\frac{2}{4}$ of line #2; ∴ $\frac{1}{2} = \frac{2}{4}$. $\frac{1}{2}$ and $\frac{2}{4}$ are *equivalent* fractions. They "look" different, but they have the same value.

Quite often when dealing with fractions we are concerned with "building" equivalent fractions. For example: Build a fraction which is equivalent to $\frac{7}{8}$ but has a denominator of 24.

GIVEN $\frac{7}{8} = \frac{?}{24}$ (We are trying to find the missing numerator.)

Solution **STEP 1** Divide the "old" denominator (8) into the "new" denominator (24).

$$\frac{7}{8} = \frac{?}{24} \qquad 8\overline{)24}^{3 \leftarrow \text{quotient}}$$

STEP 2 Multiply the quotient from Step 1 (3) by the "old" numerator (7). The product is the missing numerator.

$$\frac{7}{8} = \frac{?}{24}$$

↓ quotient from Step 1
$7 \times 3 = 21$ ← "new" numerator
↑ "old" numerator

$$\therefore \frac{7}{8} = \frac{21}{24}$$

What we've done here is to change the fraction $\frac{7}{8}$ into higher terms. We *increased* the denominator, 8, to 24, *and increased* the numerator, 7, the same number of times to 21:

$$\frac{7 \times (3)}{8 \times (3)} = \frac{21}{24}$$

We did not change the "value" of the fraction $\frac{7}{8}$ we merely changed it into an *equivalent* fraction, $\frac{21}{24}$ in higher terms (higher meaning larger denominator and numerator).

GIVEN $\quad \frac{9}{13} = \frac{?}{52}$

Solution $\quad \frac{9}{13} = \frac{?}{52}$ **STEP 1:** Thought Process: 13 goes into 52, 4 times; therefore, 4 times 9 equals 36.

$\frac{9}{13} = \frac{36}{52}$ **STEP 2:** $\frac{9 \times (4)}{13 \times (4)} = \frac{36}{52}$

$\therefore \frac{9}{13} = \frac{36}{52}$

GIVEN $\quad \frac{21}{20} = \frac{?}{100}$ (read—twenty-one twentieths equals how many hundredths?)

Solution $\quad \frac{21}{20} = \frac{?}{100}$ **STEP 1:** Thought Process: 20 goes into 100, 5 times; therefore, 5 times 21 equals 105.

$\frac{21}{20} = \frac{105}{100}$

$\therefore \frac{21}{20} = \frac{105}{100}$

FRACTIONS I: ADDITION AND SUBTRACTION

Keep in mind that as long as we increase the numerator and denominator of our "old" fraction the same number of times, then we will always end up with an equivalent fraction.

$$\frac{21 \, (\times 5)}{20 \, (\times 5)} = \frac{105}{100}$$

We increased the "old" numerator five times and we increased the "old" denominator five times ∴ we have an equivalent fraction equaling $\frac{105}{100}$.

Do you remember that multiplying any number by one gives you that number back? For example: $5 \times 1 = 5$; $100 \times 1 = 100$; $358 \times 1 = 358$. This is also true when we multiply fractions by one. This explains how we have been able to make equivalent fractions (same *value*, different numbers). Let's look at our last example again.

$$\frac{21}{20} = \frac{?}{100}$$

We determined by dividing 20 into 100 that we must have multiplied our "old" denominator, 20, by 5 to get our "new" denominator, 100. Then we multiplied our "old" numerator, 21, by the same number 5, to get our "new" numerator 105.

$$\frac{21 \times (5)}{20 \times (5)} = \frac{105}{100}$$

Taking a closer look at $\frac{\times (5)}{\times (5)}$ we see that $\frac{5}{5}$ is really **1**. So we are really multiplying $\frac{21}{20}$ by 1, which doesn't change the *value* of $\frac{21}{20}$ only its "looks."

GIVEN $\quad \dfrac{3}{5} = \dfrac{?}{45}$

Thought Process: What did we multiply 5 by in order to get 45? Well, 45 divided by 5 is 9. So, we used 9 to get to 45. That means we multiply 3 by 9 also, to get our missing numerator.

Solution $\quad \dfrac{3}{5} = \dfrac{?}{45}$

$\qquad\qquad \dfrac{3}{5} \times \left(\dfrac{9}{9}\right) = \dfrac{27}{45}$

GIVEN $\quad \dfrac{9}{16} = \dfrac{?}{96}$

Solution $\quad \dfrac{9}{16} \times \left(\dfrac{6}{6}\right) = \dfrac{54}{96}$

Thought Process: 96 divided by 16 is equal to 6, therefore, 9 times 6 is equal to 54.

EXERCISE 4-5

"Build" equivalent fractions (find the missing numerators).

*1. $\dfrac{3}{8} = \dfrac{?}{40}$

2. $\dfrac{5}{9} = \dfrac{?}{54}$

3. $\dfrac{12}{13} = \dfrac{?}{39}$

4. $\dfrac{6}{7} = \dfrac{?}{49}$

5. $\dfrac{10}{17} = \dfrac{?}{34}$

6. $\dfrac{20}{21} = \dfrac{?}{84}$

7. $\dfrac{27}{28} = \dfrac{?}{84}$

*8. $\dfrac{1}{15} = \dfrac{?}{105}$

9. $\dfrac{3}{7} = \dfrac{?}{105}$

10. $\dfrac{8}{9} = \dfrac{?}{72}$

11. $\dfrac{17}{18} = \dfrac{?}{72}$

12. $\dfrac{3}{2} = \dfrac{?}{14}$

*13. $\dfrac{6}{1} = \dfrac{?}{18}$

14. $\dfrac{9}{8} = \dfrac{?}{24}$

15. $\dfrac{10}{7} = \dfrac{?}{42}$

16. $\dfrac{100}{91} = \dfrac{?}{364}$

17. $\dfrac{82}{81} = \dfrac{?}{486}$

*18. $9 = \dfrac{?}{36}$

19. $13 = \dfrac{?}{39}$

20. $24 = \dfrac{?}{2}$

*21. $\dfrac{3}{8} = \dfrac{?}{8}$

22. $\dfrac{20}{3} = \dfrac{?}{6}$

23. $\dfrac{18}{1} = \dfrac{?}{1}$

24. $\dfrac{56}{55} = \dfrac{?}{55}$

*25. $1 = \dfrac{?}{4}$

NOTE Don't forget to check your answers before going on.

FRACTIONS I: ADDITION AND SUBTRACTION   **123**

SECTION 4-6 FRACTION ADDITION

Adding Fractions with the Same Denominator

STEP 1 To add fractions with the same denominator, add the numerators and place the sum over the common denominator.

STEP 2 Reduce your answer to simplest form.

GIVEN Add: $\dfrac{2}{9} + \dfrac{4}{9}$

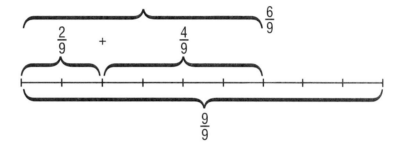

Solution

$$\begin{array}{r}\dfrac{2}{9}\\[2pt]+\,\dfrac{4}{9}\\[2pt]\hline \dfrac{6}{9} = \dfrac{2}{3}\end{array}$$

This line shown above should help you visualize what's taking place when we add $\dfrac{2}{9}$ and $\dfrac{4}{9}$ together.

GIVEN Add: $\dfrac{3}{25} + \dfrac{21}{25}$

Solution

$$\begin{array}{r}\dfrac{3}{25}\\[2pt]+\,\dfrac{21}{25}\\[2pt]\hline \dfrac{24}{25}\end{array}$$

STEP 1 AND 2

GIVEN Add: $\dfrac{15}{16} + \dfrac{13}{16}$

Solution

$$\begin{array}{r}\dfrac{15}{16}\\[2pt]+\,\dfrac{13}{16}\\[2pt]\hline \dfrac{28}{16} = \dfrac{7}{4} = 1\dfrac{3}{4}\end{array}$$

STEP 1 AND 2   Since $\dfrac{7}{4}$ is an improper fraction, we can also write its mixed number form as the answer.

Adding Fractions with Unlike Denominators

STEP 1 Find the LCM (Lowest Common Multiple—see 3-5) of the denominators. LCM of the denominators is known as the *lowest common denominator* (LCD).

STEP 2 "Build" equivalent fractions using the *LCD*. (see 4-5)

STEP 3 Add the numerators and place the sum over the common denominator.

STEP 4 Reduce.

GIVEN Add: $\dfrac{8}{21} + \dfrac{16}{35}$

Solution

$\dfrac{8}{21} = \dfrac{?}{\text{LCD}}$

$+\dfrac{16}{35} = \dfrac{?}{\text{LCD}}$

STEP 1: Find LCM for 21 and 35. LCM for 21 and 35 = **105** = **LCD**.

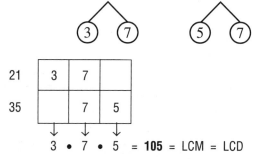

$3 \cdot 7 \cdot 5 = \mathbf{105} = \text{LCM} = \text{LCD}$

$\dfrac{8}{21} = \dfrac{40}{105}$

$+\dfrac{16}{35} = \dfrac{48}{105}$

$\dfrac{88}{105}$

STEP 2: "Build" equivalent fractions (4-5) using LCD

$\dfrac{8}{21} = \dfrac{?}{105}$ $(105 \div 21 = 5 \therefore 8 \times 5 = 40)$ $\dfrac{8}{21} = \dfrac{40}{105}$

$\dfrac{16}{35} = \dfrac{?}{105}$ $(105 \div 35 = 3 \therefore 16 \times 3 = 48)$ $\dfrac{16}{35} = \dfrac{48}{105}$

STEP 3: Add numerators and place sum over LCD.

STEP 4: Reduce. Since $\dfrac{88}{105}$ does not reduce and is not improper:

$\therefore \dfrac{8}{21} + \dfrac{16}{35} = \dfrac{88}{105}$

FRACTIONS I: ADDITION AND SUBTRACTION **125**

GIVEN Add: $\frac{13}{14} + \frac{11}{12} + \frac{3}{8}$

Solution

$\frac{13}{14} = \frac{?}{\text{LCD}}$

$\frac{11}{12} = \frac{?}{\text{LCD}}$

$+ \frac{3}{8} = \frac{?}{\text{LCD}}$

STEP 1: Find LCM for the denominators: 14, 12, 8.
LCM for 14, 12, and 8 = **168 = LCD**.

 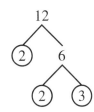 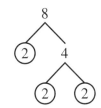

14	2	7			
12	2		2	3	
8	2		2		2

↓ ↓ ↓ ↓ ↓
2 • 7 • 2 • 3 • 2 = **168** = LCM = LCD

$\frac{13}{14} = \frac{156}{168}$

$\frac{11}{12} = \frac{154}{168}$

$+ \frac{3}{8} = \frac{63}{168}$

$\overline{\frac{373}{168}}$

STEP 2: "Build" equivalent fractions using LCD

$\frac{13}{14} = \frac{?}{168}$ (168 ÷ 14 = 12 ∴ 13 × 12 = 156) $\frac{13}{14} = \frac{156}{168}$

$\frac{11}{12} = \frac{?}{168}$ (168 ÷ 12 = 14 ∴ 11 × 14 = 154) $\frac{11}{12} = \frac{154}{168}$

$\frac{3}{8} = \frac{?}{168}$ (168 ÷ 8 = 21 ∴ 3 × 21 = 63) $\frac{3}{8} = \frac{63}{168}$

STEP 3: Add numerators and place sum over LCD.

∴ $\frac{373}{168} = 2\frac{37}{168}$

STEP 4: Reduce and change improper fraction into a mixed number.

GIVEN Add: $\frac{3}{4} + \frac{2}{8} + \frac{4}{5}$

Solution

$\frac{3}{4} = \frac{?}{\text{LCD}}$

$\frac{2}{8} = \frac{?}{\text{LCD}}$

$+ \frac{4}{5} = \frac{?}{\text{LCD}}$

STEP 1: Find LCM for the denominators: 4, 8, and 5.
LCM for 4, 8, and 5 = **40 = LCD**.

 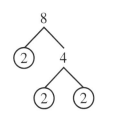

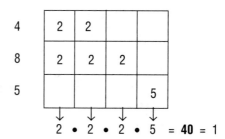

$$\frac{3}{4} = \frac{30}{40}$$

$$\frac{2}{8} = \frac{10}{40}$$

$$+\frac{4}{5} = \frac{32}{40}$$

$$\frac{72}{40} = \frac{9}{5} = 1\frac{4}{5}$$

STEP 2: "Build" equivalent fractions using LCD

STEP 3: Add.

STEP 4: Reduce and change into a mixed number.

GIVEN $\frac{2}{3}$ of an inch of rain fell in June, $\frac{3}{4}$ inch in July and $\frac{9}{10}$ inch in August. What was the total rainfall for the three months?

Solution

1. Read problem again *and* again.

2. What are we being asked to find? (total rainfall)

3. What do we know? $\frac{2}{3} + \frac{3}{4} + \frac{9}{10}$ = total rainfall in inches.

4. Do it!

$$\frac{2}{3} = \frac{40}{60}$$

$$\frac{3}{4} = \frac{45}{60}$$

$$+\frac{9}{10} = \frac{54}{60}$$

$$\frac{139}{60} = 2\frac{19}{60} \text{ inches}$$

∴ 5. Summation: a total of $2\frac{19}{60}$ inches of rain fell in June, July and August.

EXERCISE 4-6

Reduce your answers to lowest terms. Also change any improper fractions into mixed numbers.

1. $\dfrac{2}{9} + \dfrac{4}{9} =$

2. $\dfrac{3}{8} + \dfrac{7}{8} =$

*3. $\dfrac{4}{15} + \dfrac{7}{15} + \dfrac{11}{15} =$

*4. $\dfrac{3}{8} + \dfrac{4}{9} =$

5. $\dfrac{2}{7} + \dfrac{6}{9} =$

6. $\dfrac{2}{3} + \dfrac{3}{4} =$

7. $\dfrac{30}{41} + \dfrac{81}{82} =$

8. $\dfrac{5}{13} + \dfrac{31}{39} =$

9. $\dfrac{7}{11} + \dfrac{8}{33} + \dfrac{5}{12} =$

10. $\dfrac{9}{10} + \dfrac{14}{25} + \dfrac{17}{30} =$

11. $\dfrac{1}{18} + \dfrac{6}{7} + \dfrac{41}{42} =$

12. $\dfrac{3}{5} + \dfrac{8}{15} + \dfrac{17}{24} =$

13. $\dfrac{3}{14} + \dfrac{19}{21} + \dfrac{1}{5} =$

14. $\dfrac{100}{101} + \dfrac{101}{202} + \dfrac{1}{2} =$

15. $\dfrac{2}{13} + \dfrac{25}{39} + \dfrac{6}{7} =$

16. $\dfrac{200}{313} + \dfrac{301}{313} + \dfrac{85}{313} =$

17. $\dfrac{3}{4} + \dfrac{5}{8} + \dfrac{15}{16} =$

18. $\dfrac{7}{19} + \dfrac{8}{11} + \dfrac{2}{3} =$

19. $\dfrac{8}{9} + \dfrac{5}{6} + \dfrac{5}{18} =$

20. $\dfrac{1}{2} + \dfrac{13}{15} + \dfrac{2}{17} =$

21. During one terrible Winter Semester, a chemistry class had $\frac{1}{4}$ of the students drop due to bad weather and another $\frac{1}{8}$ drop due to illness. What was the total portion of the students that dropped the class?

22. On a set of steel framing plans, you are to draw a $\frac{5}{8}$ inch diameter circle to represent the hole that an appropriate sized bolt will soon fill. For some reason, the designer decides to increase the diameter by $\frac{3}{16}$ inch. What is the new diameter of the circle to be drawn?

SECTION 4-7 ADDITION OF MIXED NUMBERS

GIVEN Add: $3\frac{1}{8} + 4\frac{5}{8}$

Solution

$$3\frac{1}{8} = 3 + \frac{1}{8}$$
$$+\ 4\frac{5}{8} = +\ 4 + \frac{5}{8}$$

STEP 1, 2, AND 3 Since $\frac{1}{8}$ and $\frac{5}{8}$ have the same denominator, we do not need to perform Step 1 (find LCD) or Step 2 (build equivalent fractions). We simply add the fraction portions of our mixed numbers together as we have done previously. Plus, we will add together the whole number portions of our mixed numbers.

$$3 + \frac{1}{8}$$
$$+\ 4 + \frac{5}{8}$$
$$\overline{7 + \frac{6}{8}}$$

STEP 4 Check the fraction portion of your answer to see if it can be reduced to lower terms and/or to a mixed number.

$$7 + \frac{6}{8} = 7 + \frac{3}{4} = 7\frac{3}{4}$$

GIVEN $5\frac{2}{9} + 4\frac{7}{9} + 8\frac{5}{9}$

Solution

$$5 + \frac{2}{9}$$
$$4 + \frac{7}{9}$$
$$+\ 8 + \frac{5}{9}$$
$$\overline{17 + \frac{14}{9}}$$

STEP 1, 2, AND 3: All of the fractions have the same denominator. ∴ we add them as before. Plus we add the whole numbers together.

FRACTIONS I: ADDITION AND SUBTRACTION 129

$$17 + \frac{14}{9}$$

STEP 4: Since $\frac{14}{9}$ is an improper fraction, we must change it into mixed number form, then add it to our whole number.

$$\therefore 17 + \frac{14}{9} = 17 + \left(\frac{14}{9} = 1\frac{5}{9}\right) = 17 + 1\frac{5}{9} = 18\frac{5}{9}$$

GIVEN $10\frac{5}{18} + 2\frac{13}{18} + \frac{6}{18}$

Solution

$$10 + \frac{5}{18}$$
$$2 + \frac{13}{18}$$
$$+ \frac{6}{18}$$
$$\overline{12 + \frac{24}{18}}$$

STEP 1, 2, AND 3: All of the fractions have the same denominator, $\therefore$ add them together as before. Plus add together the whole numbers.

STEP 4: Since $\frac{24}{18}$ is an improper fraction, change to mixed number and *reduce*. Then add to whole number.

$$\therefore 12 + \frac{24}{18} = 12 + \left(\frac{24}{18} = 1\frac{6}{18} = 1\frac{1}{3}\right) = 12 + 1\frac{1}{3} = 13\frac{1}{3}$$

GIVEN Add: $2\frac{3}{8} + 4\frac{5}{6} + \frac{8}{9}$

Solution

$$2 + \frac{3}{8}$$
$$4 + \frac{5}{6}$$
$$+ \frac{8}{9}$$

This time we do not have the same denominators. $\therefore$ we must follow the procedure for adding fractions with unlike denominators. So, for the time being, ignore the whole numbers—2 and 4—and just work with the fractions:

$$\frac{3}{8} = \frac{?}{\text{LCD}}$$

$$\frac{5}{6} = \frac{?}{\text{LCD}}$$

$$\frac{8}{9} = \frac{?}{\text{LCD}}$$

$$2 + \frac{3}{8} = 2 + \frac{27}{72}$$

$$4 + \frac{5}{6} = 4 + \frac{60}{72}$$

$$\frac{8}{9} = \frac{64}{72}$$

$$6 + \frac{151}{72}$$

STEP 1 AND 2: LCD for 8, 6, and 9 is 72.

$$\therefore \frac{3}{8} = \frac{27}{72}; \quad \frac{5}{6} = \frac{60}{72}; \quad \frac{8}{9} = \frac{64}{72}.$$

STEP 3: Now add as before. This is where the whole numbers come back in.

$$\therefore 6 + \frac{151}{72} = 6 + \left(\frac{151}{72} = 2\frac{7}{72}\right) = 6 + 2\frac{7}{72} = 8\frac{7}{72}$$

STEP 4: Reduce.

GIVEN Add: $20\frac{14}{19} + 13\frac{35}{38} + 4$

Solution

$$20 + \frac{14}{19} = 20 + \frac{28}{38}$$

$$13 + \frac{35}{38} = 13 + \frac{35}{38}$$

$$+ \; 4 \qquad = \; + 4$$

$$37 + \frac{63}{38}$$

STEP 1 AND 2: LCD = 38

$$\therefore \frac{14}{19} = \frac{28}{38}; \quad \frac{35}{38} = \frac{35}{38}.$$

STEP 3: Add as before.

$$\therefore 37 + \frac{63}{38} = 37 + \left(\frac{63}{38} = 1\frac{25}{38}\right) = 37 + 1\frac{25}{38} = 38\frac{25}{38}$$

STEP 4: Reduce.

GIVEN Add: $6\frac{3}{16} + 7\frac{5}{8} + 3\frac{1}{2}$

Solution

$$6 + \frac{3}{16} = 6 + \frac{3}{16}$$

$$7 + \frac{5}{8} = 7 + \frac{10}{16}$$

$$3 + \frac{1}{2} = 3 + \frac{8}{16}$$

$$16 + \frac{21}{16}$$

STEP 1, 2, 3, 4: LCD = 16

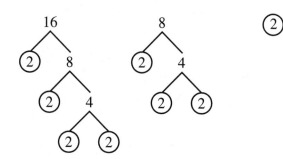

$$\therefore 16 + \frac{21}{16} = 16 + \left(\frac{21}{16} = 1\frac{5}{16}\right) = 16 + 1\frac{5}{16} = 17\frac{5}{16}$$

EXERCISE 4-7

Reduce all answers.

*1. $5\frac{1}{2} + 4\frac{1}{2} + \frac{1}{2} =$

2. $16\frac{3}{8} + 10\frac{5}{8} + 3\frac{7}{8} =$

3. $15\frac{9}{10} + \frac{7}{10} + 2\frac{1}{10} =$

*4. $7 + 8\frac{1}{3} =$

5. $9\frac{5}{12} + 20 + 1\frac{11}{12} =$

6. $2\frac{4}{5} + 6\frac{9}{20} =$

*7. $23\frac{7}{18} + \frac{4}{9} + 8 =$

8. $4\frac{1}{20} + 3\frac{7}{11} + 5\frac{9}{44} =$

9. $60\frac{1}{2} + 43\frac{2}{3} + 1 =$

10. $13\frac{15}{16} + 2\frac{13}{20} + \frac{3}{4} =$

11. $3 + 24\frac{1}{7} + 14\frac{7}{9} =$

*12. $155\frac{13}{24} + 260 + \frac{5}{8} =$

13. $6\frac{2}{15} + 32\frac{5}{9} + 8\frac{5}{6} =$

14. $2\frac{3}{4} + 41\frac{1}{8} + 29\frac{7}{16} =$

15. $8\frac{120}{121} + 9\frac{6}{11} + 14\frac{1}{121} =$

16. $28\frac{5}{8} + 3\frac{4}{9} + 4\frac{5}{14} + \frac{3}{8} =$

17. $10 + 13\frac{2}{3} + \frac{15}{16} + 7\frac{7}{9} =$

18. $3\frac{1}{3} + 14\frac{2}{3} + 60 + 7\frac{7}{9} =$

19. $\frac{23}{25} + 5\frac{19}{20} + 14\frac{11}{30} + 32 =$

*20. $67\frac{4}{5} + \frac{1}{2} + 8\frac{16}{17} + 123\frac{7}{34} =$

*21. An alloy contains a mixture of $4\frac{2}{3}$ ounces of silver, $1\frac{3}{8}$ ounces of gold, and $\frac{5}{15}$ ounce of platinum. What is the weight of the alloy?

*22. What is the total horsepower available if five motors have the following power ratings:

$7\frac{1}{2}$ hp, $4\frac{1}{6}$ hp, $5\frac{2}{3}$ hp, $2\frac{1}{5}$ hp?

SECTION 4-8 FRACTION SUBTRACTION

▣ Procedure for Subtracting Fractions

STEP 1   Find LCD.

STEP 2   "Build" equivalent fractions using *LCD*.

STEP 3   Subtract the numerators and place the difference over the LCD.

STEP 4   Reduce.

GIVEN $\dfrac{5}{9} - \dfrac{2}{9}$ Step 1 and 2 can be skipped since the fractions have the same denominator.

Solution

$$\begin{array}{r} \dfrac{5}{9} \\ -\dfrac{2}{9} \\ \hline \dfrac{3}{9} \end{array} \leftarrow \text{difference}$$

STEP 3: Subtract numerators (5 − 2 = 3) and place difference over LCD.

$$\dfrac{3}{9} = \dfrac{1}{3}$$

STEP 4: Reduce.

▬ PICTURE THIS ▬

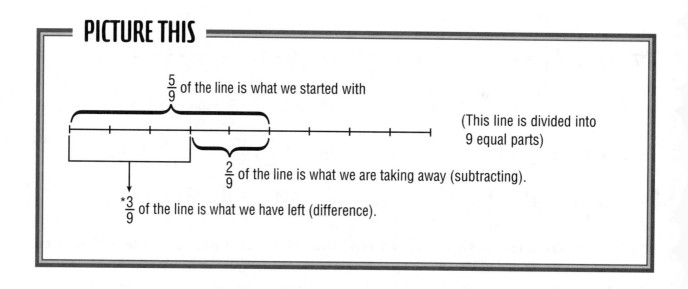

$\dfrac{5}{9}$ of the line is what we started with

(This line is divided into 9 equal parts)

$\dfrac{2}{9}$ of the line is what we are taking away (subtracting).

*$\dfrac{3}{9}$ of the line is what we have left (difference).

FRACTIONS I: ADDITION AND SUBTRACTION 133

GIVEN Subtract: $\frac{10}{17}$ from $\frac{15}{17}$

Solution

$\frac{15}{17}$

Be careful! Commutative Property does *not* hold for subtraction. We must be very careful which fraction is placed first.

$-\frac{10}{17}$

$\frac{15}{17}$

Steps 1 and 2 can be skipped.

$-\frac{10}{17}$

STEP 3: Subtract (15 − 10 = 5). Place difference over LCD.

$\frac{5}{17}$ ← difference

STEP 4: Reduce; $\frac{5}{17}$ is reduced.

GIVEN $\frac{7}{8} - \frac{2}{3}$

Solution

$\frac{7}{8} = \frac{21}{24}$

STEP 1: Find LCD. LCD for 8 and 3 is 24.

$-\frac{2}{3} = \frac{16}{24}$

STEP 2: "Build" equivalent fractions.

$\frac{5}{24}$

STEP 3: Subtract.
STEP 4: Reduce.

GIVEN $\frac{19}{20}$ minus $\frac{1}{15}$ equals what number?

Solution

$\frac{19}{20} = \frac{57}{60}$

STEP 1: LCD = 60

$-\frac{1}{15} = \frac{4}{60}$

STEP 2: Equivalent fractions.

$\frac{53}{60}$

STEP 3: Subtract.
STEP 4: Reduce.

EXERCISE 4-8

*1. $\dfrac{3}{8} - \dfrac{1}{8} =$

2. $\dfrac{4}{5} - \dfrac{2}{5} =$

*3. $\dfrac{15}{16} - \dfrac{3}{16} =$

*4. $\dfrac{2}{31} - \dfrac{2}{31} =$

5. $\dfrac{61}{75} - \dfrac{11}{75} =$

*6. $\dfrac{3}{4} - \dfrac{1}{3} =$

7. $\dfrac{2}{9} - \dfrac{1}{7} =$

8. $\dfrac{5}{6} - \dfrac{1}{10} =$

9. $\dfrac{20}{21} - \dfrac{3}{16} =$

10. $\dfrac{18}{35} - \dfrac{4}{21} =$

*11. Subtract $\dfrac{9}{28}$ from $\dfrac{13}{14}$.

*12. $\dfrac{38}{45}$ minus $\dfrac{7}{15}$ equals what number?

13. Subtract $\dfrac{2}{11}$ from $\dfrac{2}{3}$.

14. Subtract $\dfrac{5}{9}$ from $\dfrac{41}{42}$.

*15. What does $\dfrac{3}{16}$ minus $\dfrac{1}{8}$ equal?

*16. You are to administer $\frac{2}{3}$ teaspoon of medication to a patient; however, the doctor has just given instructions to decrease the amount of medication given by $\frac{1}{4}$ teaspoon. How much should the patient receive?

*17. $\frac{5}{9}$ of the total floor space of a clothing store is used for displaying merchanddise, cash register area, and dressing rooms. If $\frac{3}{10}$ of the total floor space is used for dressing rooms alone, how much of the total floor space is left for displaying merchandise and cash register area?

SECTION 4-9 MIXED NUMBER SUBTRACTION AND BORROWING

Procedure for Subtraction of Mixed Number Fractions

STEP 1 Find LCD.

STEP 2 "Build" equivalent fractions.

STEP 3 Perform "borrowing" procedure if necessary.

STEP 4 Subtract numerators of fractions and place difference over LCD, and subtract whole numbers.

STEP 5 Reduce fraction portion of mixed number, if possible.

> **NOTE** Step 3 ("borrow" if necessary) will be explained later in this section. The first two examples that follow immediately do not require this step.

GIVEN $4\frac{2}{3} - 2\frac{1}{2}$

Solution

$$4\frac{2}{3} = 4\frac{4}{6}$$
$$-2\frac{1}{2} = 2\frac{3}{6}$$
$$\overline{\phantom{-2\frac{1}{2}=\;}2\frac{1}{6}}$$

STEP 1: Find LCD. LCD of 3 and 2 equals 6.
STEP 2: Build equivalent fractions.
STEP 3: Not required for this problem.
STEP 4: Subtract fractions ($\frac{4}{6} - \frac{3}{6} = \frac{1}{6}$)
Subtract whole numbers ($4 - 2 = 2$)
STEP 5: Reduce if possible.

GIVEN $10\frac{8}{9} - 3\frac{4}{5}$

Solution

$$10\frac{8}{9} = 10\frac{40}{45}$$
$$-3\frac{4}{5} = 3\frac{36}{45}$$
$$\overline{\phantom{-3\frac{4}{5}=\;}7\frac{4}{45}}$$

STEP 1: LCD of 9 and 5 is 45.
STEP 2: Equivalent fractions.
STEP 3: Not required.
STEP 4: Subtract fractions and subtract whole numbers.
STEP 5: Reduce if possible.

GIVEN $\quad 9\frac{1}{3} - 2\frac{2}{3}$

Solution $\quad 9\frac{1}{3} \quad 1 - 2 = ?$ **STEPS 1 AND 2:** Same denominators, ∴ skip these steps.

$\quad\quad\quad -2\frac{2}{3}$ **STEP 3:** Since we cannot subtract $\frac{2}{3}$ from $\frac{1}{3}$ ($\frac{1}{3} - \frac{2}{3} = ?$) we must now perform the borrowing step.

Borrowing

Our procedure for borrowing in fractions is *similar* to borrowing in whole numbers. Let's look back at our original problem:

$$9\frac{1}{3} \quad \leftarrow \text{We need to increase this numerator in order to perform the subtraction.}$$

$$-2\frac{2}{3}$$

$$\overset{8}{\cancel{9}} + \frac{1}{3} \quad \leftarrow \text{Let's borrow a } \mathbf{1} \text{ from the whole number (9), leaving us with a whole number 8.}$$

$$-2 + \frac{2}{3}$$

$$\overset{8}{\cancel{9}} + \left(1 + \frac{1}{3}\right) \quad \leftarrow \text{The whole number } \mathbf{1} \text{ that we just borrowed must now be added to the fraction } \frac{1}{3}.$$

$$-2\frac{2}{3} \quad\quad \therefore\ 1 + \frac{1}{3} = 1\frac{1}{3} = \frac{4}{3} \quad \text{(Remember how to turn a mixed number into an improper fraction (see 4-3)?)}$$

$$\overset{8}{\cancel{9}}\frac{4}{3}$$
$$\quad\quad 4 - 2 = 2 \quad \text{Now that we have enough to subtract from, we can continue to Steps 4 and 5.}$$
$$-2\frac{2}{3}$$

$$6\frac{2}{3}$$

FRACTIONS I: ADDITION AND SUBTRACTION 137

GIVEN
$$27\frac{5}{32}$$
$$-16\frac{31}{32}$$

Solution

$$27\frac{5}{32}$$
$$-16\frac{31}{32}\bigg]\text{same}$$

STEPS 1 AND 2: Since we already have the same denominator, skip these steps.

$$27\frac{5}{32}$$
$$-16\frac{31}{32}\bigg]\ 5-31=?$$

STEP 3: Borrow, since we cannot do the subtraction *until* we borrow. Borrow a 1 from 27 then add the borrowed 1 to the fraction $\frac{5}{32}$.

$$1+\frac{5}{32}=1\frac{5}{32}=\frac{37}{32}$$

$$\overset{26}{\cancel{27}}+\left(1+\frac{5}{32}\right)=26+\left(1\frac{5}{32}=\frac{37}{32}\right)=26\frac{37}{32}\quad\text{Now we have enough to subtract from.}$$
$$-16\frac{31}{32}\qquad\qquad=16\frac{31}{32}\qquad\qquad=16\frac{31}{32}$$

$$\therefore\quad 27\frac{5}{32}\ =\ 26\frac{37}{32}$$
$$-16\frac{31}{32}=-16\frac{31}{32}$$
$$\overline{\qquad\qquad 10\frac{6}{32}=10\frac{3}{16}}$$

STEPS 4 AND 5

GIVEN
$$18\frac{9}{20}-9\frac{29}{30}$$

Solution

$$18\frac{9}{20}=18\frac{27}{60}$$
$$-9\frac{29}{30}=9\frac{58}{60}$$

STEPS 1 AND 2: LCD = 60. $\therefore\ \frac{9}{20}=\frac{27}{60};\ \frac{29}{30}=\frac{58}{60}$

$$18\frac{9}{20}=\ 18\frac{27}{60}=\ \overset{17}{\cancel{18}}\frac{87}{60}$$
$$-9\frac{29}{30}=-9\frac{58}{60}=-9\frac{58}{60}$$
$$\overline{\qquad\qquad\qquad\qquad\qquad 8\frac{29}{60}}$$

STEP 3: Borrow. $1+\frac{27}{60}=1\frac{27}{60}=\frac{87}{60}$

STEPS 4, 5, AND 6: Subtract and reduce.

GIVEN $10 - 3\frac{2}{5}$

Solution

10
$- 3\frac{2}{5}$

Since 10 has no fraction portion and we *do* need a fraction to subtract $\frac{2}{5}$ from; we must skip right to Step 3.

$\cancel{10}^{9}\frac{5}{5}$
$- 3\frac{2}{5}$

STEP 3: Borrow $(1 = \frac{5}{5})$ we choose 5 as our denominator since $\frac{2}{5}$ has 5 as its denominator.

$$\therefore 10 = 9\frac{5}{5}$$

$\cancel{10}^{9}\frac{5}{5}$
$- 3\frac{2}{5}$
$\overline{6\frac{3}{5}}$

STEPS 4 AND 5

GIVEN $6\frac{1}{3}$
$- 4$

Since no fraction is being subtracted from $\frac{1}{3}$, there is *no* need to borrow.

Solution

$6\frac{1}{3}$
$- 4$
$\overline{2\frac{1}{3}}$

STEP 4: Since we are subtracting "nothing" from $\frac{1}{3}$, we bring the $\frac{1}{3}$ down as part of our answer. $(\frac{1}{3} - 0 = \frac{1}{3})$ Subtract whole numbers.

STEP 5: Reduce if possible.

GIVEN $4\frac{2}{3} + 2\frac{5}{8} - 3\frac{15}{16}$

Solution $4\frac{2}{3} + 2\frac{5}{8} = ?$

Following order of operations, we must work this problem from left to right. $\therefore$ we add $4\frac{2}{3}$ plus $2\frac{5}{8}$ together, then subtract $3\frac{15}{16}$ from their sum.

then,

$? - 3\frac{15}{16} =$ FINAL ANSWER

FRACTIONS I: ADDITION AND SUBTRACTION **139**

$$4\frac{2}{3} = 4\frac{16}{24}$$
$$+ 2\frac{5}{8} = 2\frac{15}{24}$$
$$\overline{\phantom{+ 2\frac{5}{8}}\ 6\frac{31}{24}}$$
$$\therefore 6\frac{31}{24} = 7\frac{7}{24}$$

STEP 1: LCD = 24.

STEP 2: $\frac{2}{3} = \frac{16}{24}$; $\frac{5}{8} = \frac{15}{24}$

STEP 3: For addition of mixed numbers. Add fractions and whole numbers.

$$7\frac{7}{24} - 3\frac{15}{16} = ?$$

Now we complete the problem by subtracting $3\frac{15}{16}$ from $7\frac{7}{24}$.

$$7\frac{7}{24} = 7\frac{14}{48} = 6\frac{62}{48}$$
$$- 3\frac{15}{16} = 3\frac{45}{48} = 3\frac{45}{48}$$
$$\overline{\phantom{- 3\frac{15}{16}}\ 3\frac{17}{48}}$$

STEP 1: LCD = 48.

STEP 2: $\frac{7}{24} = \frac{14}{48}$; $\frac{15}{16} = \frac{45}{48}$

STEP 3: For Subtraction of mixed numbers. Borrow.

STEP 4: Subtract fractions and whole numbers.

STEP 5: Reduce.

🔲 An Alternate Method for Adding and Subtracting Mixed Number Fractions

GIVEN $3\frac{1}{2} + 2\frac{5}{8}$

Solution

$$3\frac{1}{2} = \frac{7}{2} = \frac{28}{8}$$
$$+ 2\frac{5}{8} = \frac{21}{8} = \frac{21}{8}$$
$$\overline{\phantom{+ 2\frac{5}{8}}\ \frac{49}{8} = 6\frac{1}{8}}$$

Change the mixed numbers into improper fractions. Find the LCD and make equivalent fractions (LCD for 2 and 8 = 8). Then add the numerators and place their sum over the LCD.

Reduce.

GIVEN $\quad 5\frac{2}{9} - 3\frac{3}{4}$

Solution

$$5\frac{2}{9} = \frac{47}{9} = \frac{188}{36} \qquad \text{LCD for 9 and 4} = 36$$

$$-\,3\frac{3}{4} = \frac{15}{4} = \frac{135}{36}$$

$$\frac{53}{36} = 1\frac{17}{36}$$

This method of converting mixed numbers to improper fractions, finding the LCD, changing the improper fractions into equivalent fractions, and performing the indicated operation does have its advantages. The main advantage being the elimination of the "borrowing" process. However, when working with numbers such as $35\frac{34}{37} + 68\frac{87}{90}$, this process becomes too cumbersome, simply because the numbers we have to work with are so large. It takes us too long to turn them into improper fractions. Also the chances of making a mistake increase when working with such large numbers. The final choice of which method to use is yours. The following exercise will give you experience at working with both methods.

FRACTIONS I: ADDITION AND SUBTRACTION

EXERCISE 4-9

In problems 1–10, use both the mixed number and the improper fraction methods for subtraction. After problem 10, it's your choice.

*1. $5\frac{1}{4} - 3\frac{3}{4}$

2. $12\frac{1}{3} - 10\frac{2}{3}$

*3. $3\frac{3}{8} - \frac{7}{8}$

*4. $6 - 4\frac{5}{9}$

5. $4\frac{5}{6} - 3\frac{13}{14}$

*6. $7\frac{2}{7} - \frac{27}{28}$

7. $8\frac{6}{11} - 5\frac{2}{3}$

8. $1\frac{3}{50} - \frac{4}{5}$

9. $2\frac{7}{9} - 1\frac{17}{18}$

*10. $9\frac{9}{10} - 3$

11. $15\frac{5}{8} - 7\frac{3}{4}$

12. $22\frac{6}{11} - 13\frac{7}{9}$

*13. $41 - 26\frac{1}{2}$

14. $63\frac{5}{6} - 48\frac{2}{3}$

15. $32\frac{15}{16} - 23$

*16. $18\frac{42}{47} - 17\frac{140}{141}$

17. $89\frac{2}{3} - 17\frac{5}{6}$

18. $15\frac{6}{101} - 1\frac{1}{2}$

19. $205\frac{3}{4} - 140\frac{3}{7}$

*20. $1000 - 998\frac{23}{1042}$

*21. $8\frac{2}{3} + 6\frac{5}{9} - 3\frac{17}{18}$

22. $12\frac{1}{2} + 36\frac{5}{8} - \frac{23}{24}$

23. $6 + 3\frac{1}{10} - 5\frac{14}{15}$

*24. $4\frac{5}{9} + 16\frac{12}{27} - 10\frac{5}{8}$

25. $38\frac{2}{3} + 4\frac{3}{4} - 15$

26. $\frac{7}{16} + 5\frac{1}{9} - 3\frac{2}{3}$

*27. $43\frac{1}{12} + 25 - \frac{7}{8}$

28. $17\frac{13}{15} + \frac{3}{7} - 1\frac{6}{11}$

29. $75 + 53\frac{1}{2} - \frac{1}{2}$

*30. $306\frac{5}{6} + 208\frac{3}{13} - 367\frac{20}{24}$

*31. A carpenter cuts a block $5\frac{1}{4}$ inch long. If he then planes $\frac{3}{16}$ inch off, how long is the finished block?

*32. You have been ordered to decrease a patient's medication by $1\frac{1}{8}$ mg (mg = milligram). If the patient has previously been receiving $4\frac{3}{4}$ mg of medication, how much should he receive now?

*33. Calculate the distance around the geometric figure pictured below.

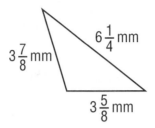

*34. A solution is prepared by mixing $2\frac{2}{3}$ quarts water; $\frac{3}{4}$ quart lemon juice; $1\frac{5}{8}$ quarts orange juice; and $2\frac{1}{2}$ quarts alcohol. If someone then proceeds to drink $2\frac{1}{4}$ quarts of this mixture, how much is left?

*35. Calculate the inside diameter of the tube shown below if the wall thickness is $\frac{3}{16}$ inch.

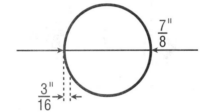

HINT: There are two wall thicknesses to account for.

YOU SHOULD NOW BE READY FOR THE CHAPTER REVIEW TEST. GOOD LUCK!!

CHAPTER 4

REVIEW TEST #1

1. $\dfrac{4}{15} + \dfrac{7}{15} - \dfrac{2}{15} = \underline{}$ (see 4-6)

2. $\dfrac{8}{13} = \dfrac{?}{39}$ (see 4-4)

3. If 6 is the numerator and 7 the denominator, the fraction is $\underline{}$ (see 4-2)

4. Write as improper fractions: (see 4-3)

   a. $5\dfrac{2}{3}$ b. $16\dfrac{1}{9}$

5. Write as mixed numbers and reduce if possible: (see 4-3 and 4-5)

   a. $\dfrac{45}{6}$ b. $\dfrac{213}{40}$

Reduce answers in 6–10 if possible.

6. $\dfrac{3}{11} + \dfrac{21}{25} + \dfrac{11}{15} = \underline{}$ (see 4-6)

7. The sum of $4\dfrac{5}{9}$, $7\dfrac{3}{8}$ and $\dfrac{5}{12}$ is $\underline{}$ (see 4-7)

8. The difference of $\dfrac{14}{19}$ and $\dfrac{27}{38}$ is $\underline{}$ (see 4-8)

9. Subtract $5\dfrac{3}{10}$ from $8\dfrac{2}{3}$.   (see 4-9)

10. $4\dfrac{1}{8} + 9\dfrac{4}{7} - 1\dfrac{41}{42} = \underline{}$ (see 4-9)

11. A patient's medication is to be increased by $2\dfrac{1}{2}$ mg. If the patient has previously been receiving $3\dfrac{1}{4}$ mg of medication, how much should he receive now?   (see 4-9)

CHECK YOUR ANSWERS If you missed three (3) or more answers, correct your mistakes, study this chapter again, and take Review Test # 2. If you missed fewer than three problems, correct your mistakes and ask your instructor for the topic quiz on Fractions I.

CHAPTER 4

Review Test #2

1. $\dfrac{3}{20} + \dfrac{17}{20} - \dfrac{11}{20} = \underline{}$

2. $\dfrac{6}{19} = \dfrac{?}{38}$

3. If 12 is the denominator and 17 is the numerator, the fraction is $\underline{}$

4. Write as improper fractions:

   a. $3\dfrac{3}{16}$ b. $20\dfrac{1}{35}$

5. Write as mixed numbers and reduce if possible:

   a. $\dfrac{16}{9}$ b. $\dfrac{420}{21}$

Reduce answers in 6–10 if possible.

6. $\dfrac{6}{7} + \dfrac{4}{21} + \dfrac{3}{14} = \underline{}$

7. The sum of $5\dfrac{4}{9}$, $13\dfrac{2}{3}$ and $\dfrac{3}{16} = \underline{}$

8. The difference of $\dfrac{23}{29}$ and $\dfrac{23}{87}$ is $\underline{}$

9. Subtract $8\dfrac{4}{5}$ from $13\dfrac{1}{8}$.

10. $2\dfrac{3}{10} + 5\dfrac{4}{21} - 2\dfrac{24}{25} = \underline{}$

11. A solution is prepared by mixing $4\dfrac{1}{3}$ pints of water, $2\dfrac{3}{4}$ pints of ginger ale, $\dfrac{1}{2}$ pint of lime juice, and 3 pints of alcohol. If you have to find a container to put this in, how many pints does the container have to hold?

CHECK YOUR **ANSWERS** — If you missed three (3) or more problems, correct your mistakes, study the sections in this chapter that gave you the most trouble, and take Review Test # 3. Otherwise, ask your instructor for the topic quiz on Fractions I.

CHAPTER 4

Review Test #3

1. $\dfrac{14}{19} + \dfrac{3}{19} - \dfrac{1}{19} =$ _____

2. $\dfrac{6}{7} = \dfrac{?}{49}$

3. If 20 is the denominator and 21 the numerator, the fraction is _____

4. Write as improper fractions:

    a. $6\dfrac{4}{7}$ b. $13\dfrac{5}{18}$

5. Write as mixed numbers and reduce if possible:

    a. $\dfrac{28}{5}$ b. $\dfrac{370}{15}$

Reduce answers in 6–10 if possible.

6. $\dfrac{3}{14} + \dfrac{9}{10} + \dfrac{31}{35} =$ _____

7. The sum of $8\dfrac{5}{11}$, $3\dfrac{2}{3}$ and $\dfrac{5}{6}$ is _____

8. The difference of $\dfrac{28}{31}$ and $\dfrac{17}{32}$ is _____

9. Subtract $2\dfrac{15}{16}$ from $5\dfrac{1}{4}$.

10. $6\dfrac{4}{7} + 12\dfrac{2}{33} - 3\dfrac{76}{77} =$ _____

11. The total distance around (perimeter) a room is $48\dfrac{2}{3}$ ft. If three of the sides are measured to be $12\dfrac{1}{8}$, $10\dfrac{5}{16}$ and $13\dfrac{1}{2}$ feet, what should the length of the last side be?

CHECK YOUR ANSWERS If you missed three (3) or more answers, ask your instructor for help as soon as possible. Otherwise, you are ready for the topic quiz on Fractions I.

CHAPTER 5

FRACTIONS II: MULTIPLICATION AND DIVISION

INTRODUCTION

In this chapter, you will become reacquainted with multiplication and division of fractions and mixed numbers.

SECTION 5-1 MULTIPLICATION OF FRACTIONS AND CANCELLATION

The nice thing about multiplying and dividing fractions is that you *don't* need a common denominator. Let's look at some multiplication problems involving fractions:

GIVEN $\quad \dfrac{3}{5} \cdot \dfrac{4}{7} \quad \leftarrow$ numerators
$\qquad\qquad\qquad\quad \leftarrow$ denominators

Solution

STEP 1	Multiply the numerators.	$3 \cdot 4 = \mathbf{12}$
STEP 2	Multiply the denominators.	$5 \cdot 7 = \mathbf{35}$
STEP 3	Write the products.	$\dfrac{12}{35}$

$$\therefore \dfrac{3}{5} \cdot \dfrac{4}{7} = \dfrac{3 \cdot 4}{5 \cdot 7} = \underline{\underline{\dfrac{12}{35}}}$$

GIVEN $\quad \dfrac{15}{19} \cdot \dfrac{1}{4}$

Solution $\quad \dfrac{15}{19} \cdot \dfrac{1}{4} = \dfrac{15 \cdot 1}{19 \cdot 4} = \dfrac{15}{76}$

STEPS 1 AND 2: Multiply the numerators together and multiply the denominators together.

$$\therefore \dfrac{15}{19} \cdot \dfrac{1}{4} = \underline{\underline{\dfrac{15}{76}}}$$

GIVEN $\quad \dfrac{3}{8} \cdot \dfrac{5}{7} \cdot \dfrac{1}{11}$

Solution $\quad \dfrac{3}{8} \cdot \dfrac{5}{7} \cdot \dfrac{1}{11} = \dfrac{3 \cdot 5 \cdot 1}{8 \cdot 7 \cdot 11} = \dfrac{15}{616}$

STEPS 1 AND 2: Multiply numerators; multiply denominators.

$$\therefore \dfrac{3}{8} \cdot \dfrac{5}{7} \cdot \dfrac{1}{11} = \underline{\underline{\dfrac{15}{616}}}$$

Before we continue on, let's look again at these last three examples. Notice that none of the answers were reducable. All of our answers were fractions in simplest form.

FRACTIONS II: MULTIPLICATION AND DIVISION   149

Now look at this example:

GIVEN $\dfrac{7}{8} \cdot \dfrac{48}{63}$

Solution $\dfrac{7}{8} \cdot \dfrac{48}{63} = \dfrac{7 \cdot 48}{8 \cdot 63} = \dfrac{336}{504}$ **STEPS 1 AND 2:** Multiply numerators; multiply denominators.

$\dfrac{336}{504} = \dfrac{168}{252} = \dfrac{84}{126} = \dfrac{42}{63} = \dfrac{14}{21} = \dfrac{2}{3}$ $\dfrac{336}{504}$ can be reduced.

FINALLY WE GET THE ANSWER $\dfrac{2}{3}$!

Was there an easier and/or faster way of getting the answer $\dfrac{2}{3}$? YES!

▣ Cancellation: A Way to Reduce Fraction Products before you Multiply

At this point, it would be a good idea to review the divisibility tests you learned in Chapter 3, Section 2. You will use these tests many times. Let's look at our previous problem again:

GIVEN $\dfrac{7}{8} \cdot \dfrac{48}{63}$

Solution   Using cancellation:

STEP 1   Look for common factors (see 2-3) in the numerators *and* denominators. Since both 7 (numerator) and 63 (denominator) can be divided evenly by 7 (7 is a common factor for both 7 and 63), let's use a 7 to cancel or reduce the numerator *and* denominator of our problem.

$\dfrac{7}{8} \cdot \dfrac{48}{63}$

$(7 \div 7 = 1) \rightarrow \dfrac{\cancel{7}^{1}}{8} \cdot \dfrac{48}{\cancel{63}_{9}} \leftarrow (63 \div 7 = 9)$

STEP 2   Look for more common factors in the numerators *and* denominators. Since 48 and 8 share 8 as a common factor, let's reduce the numerator (48) and the denominator (8) of our problem by dividing both by 8.

$(8 \div 8 = 1) \rightarrow \dfrac{\cancel{7}^{1}}{\cancel{8}_{1}} \cdot \dfrac{\cancel{48}^{6}}{\cancel{63}_{9}} \leftarrow (48 \div 8 = 6)$

STEP 3 Look for more common factors shared by a numerator *and* denominator. Since the remaining numerator (6) and denominator (9) share 3 as a common factor, let's further reduce the numerator and denominator by dividing both by 3.

$$\frac{\cancel{7}}{\cancel{8}} \cdot \frac{\cancel{48}^{\,2}}{\cancel{63}_{\,3}} \quad \begin{array}{l}\leftarrow (6 \div 3 = 2) \\ \\ \leftarrow (9 \div 3 = 3)\end{array}$$

(with 1's from prior cancellations on 7/8 and 48/63)

STEP 4 If there are no more common factors shared by a numerator and denominator, then multiply the remaining numerators together and then multiply the remaining denominators together.

$$\frac{\cancel{7}}{\cancel{8}} \cdot \frac{\cancel{48}}{\cancel{63}} = \frac{1 \cdot 2}{1 \cdot 3} = \underline{\underline{\frac{2}{3}}}$$

IMPORTANT: Cancellation can only take place in *multiplication*. Also cancellation is done by finding common factors shared by a *numerator* and a *denominator*.

You *never* cancel a numerator with a numerator or a denominator with a denominator.

GIVEN $\quad \dfrac{4}{5} \cdot \dfrac{8}{25}$

Solution $\quad$ Be Careful!

$\dfrac{4}{5} \cdot \dfrac{8}{25}$ $\qquad$ **STEP 1:** Look for common factors shared by a numerator *and* denominator. *There are none.*

$\therefore \dfrac{4}{5} \cdot \dfrac{8}{25} = \underline{\underline{\dfrac{32}{125}}}$ $\qquad$ **STEP 2:** Multiply numerators and multiply denominators.

GIVEN $\quad \dfrac{15}{36} \cdot \dfrac{18}{40}$

Solution $\quad \dfrac{15}{36} \cdot \dfrac{18}{40}$ $\qquad$ **STEP 1:** Look for common factors shared by a numerator *and* denominator. 15 (numerator) and 40 (denominator) share 5 as a common factor.

$(15 \div 5 = 3) \rightarrow \dfrac{\cancel{15}^{\,3}}{36} \cdot \dfrac{18}{\cancel{40}_{\,8}} \leftarrow (40 \div 5 = 8)$

$$(36 \div 9 = 4) \rightarrow \frac{\overset{3}{\cancel{15}}}{\underset{4}{\cancel{36}}} \cdot \frac{\overset{2}{\cancel{18}}}{\underset{8}{\cancel{40}}} \leftarrow (18 \div 9 = 2)$$

STEP 2: More cancellations? YES! 18 and 36 share 9 as a common factor.

$$ \frac{\overset{3}{\cancel{15}}}{\underset{\underset{2}{\cancel{4}}}{\cancel{36}}} \cdot \frac{\overset{\overset{1}{\cancel{2}}}{\cancel{18}}}{\underset{8}{\cancel{40}}} \leftarrow (2 \div 2 = 1)$$
$$(4 \div 2 = 2) \rightarrow$$

STEP 3: More cancellation? YES. Notice we can either divide the numerator 2 and denominator 4 by their common factor, *or* we can divide the numerator 2 and denominator 8 by their common factor. We *cannot* do *both*, however.

$$\frac{\overset{3}{\cancel{15}}}{\underset{\underset{2}{\cancel{4}}}{\cancel{36}}} \cdot \frac{\overset{\overset{1}{\cancel{2}}}{\cancel{18}}}{\underset{8}{\cancel{40}}} = \frac{3 \cdot 1}{2 \cdot 8} = \frac{3}{\underline{\underline{16}}}$$

STEP 4: More cancellation? NO. Multiply remaining numerators together and denominators together.

GIVEN

$$\frac{3}{16} \cdot \frac{24}{18} \cdot \frac{4}{9}$$

Solution

$$\frac{\overset{1}{\cancel{3}}}{16} \cdot \frac{24}{\underset{6}{\cancel{18}}} \cdot \frac{4}{9}$$

STEP 1: Look for all possible cancellations.

$$\frac{\overset{1}{\cancel{3}}}{\underset{4}{\cancel{16}}} \cdot \frac{\overset{6}{\cancel{24}}}{\underset{6}{\cancel{18}}} \cdot \frac{4}{9}$$

NOTE: For the purpose of showing how the cancellation is being done, the problem will be re-written each time another cancellation is done.

$$\frac{\overset{1}{\cancel{3}}}{\underset{4}{\cancel{16}}} \cdot \frac{\overset{\overset{1}{\cancel{6}}}{\cancel{24}}}{\underset{\underset{1}{\cancel{6}}}{\cancel{18}}} \cdot \frac{4}{9}$$

$$\frac{\overset{1}{\cancel{3}}}{\underset{\underset{1}{\cancel{4}}}{\cancel{16}}} \cdot \frac{\overset{\overset{1}{\cancel{6}}}{\cancel{24}}}{\underset{\underset{1}{\cancel{6}}}{\cancel{18}}} \cdot \frac{\overset{1}{\cancel{4}}}{9} = \frac{1 \cdot 1 \cdot 1}{1 \cdot 1 \cdot 9} = \frac{1}{9}$$

However, this is what we exepct to see.

$$\therefore \frac{3}{16} \cdot \frac{24}{18} \cdot \frac{4}{9} = \frac{1}{\underline{\underline{9}}}$$

152 CHAPTER 5

GIVEN $\dfrac{21}{32} \cdot \dfrac{6}{15} \cdot \dfrac{16}{28}$

Solution $\dfrac{\overset{7}{\cancel{21}}}{32} \cdot \dfrac{6}{\underset{5}{\cancel{15}}} \cdot \dfrac{16}{28}$ **STEP 1:** Look for all possible cancellations.

$\dfrac{\overset{7}{\cancel{21}}}{\underset{16}{\cancel{32}}} \cdot \dfrac{\overset{3}{\cancel{6}}}{\underset{5}{\cancel{15}}} \cdot \dfrac{16}{28}$

$\dfrac{\overset{7}{\cancel{21}}}{\underset{\underset{1}{16}}{\cancel{32}}} \cdot \dfrac{\overset{3}{\cancel{6}}}{\underset{5}{\cancel{15}}} \cdot \dfrac{\overset{1}{\cancel{16}}}{28}$

$\dfrac{\overset{\overset{1}{\cancel{7}}}{\cancel{21}}}{\underset{\underset{1}{\cancel{16}}}{\cancel{32}}} \cdot \dfrac{\overset{3}{\cancel{6}}}{\underset{5}{\cancel{15}}} \cdot \dfrac{\overset{1}{\cancel{16}}}{\underset{4}{\cancel{28}}} = \dfrac{1 \cdot 3 \cdot 1}{1 \cdot 5 \cdot 4} = \dfrac{3}{20}$ This line is what we expect to see.

$\therefore \dfrac{21}{32} \cdot \dfrac{6}{15} \cdot \dfrac{16}{28} = \underline{\underline{\dfrac{3}{20}}}$

GIVEN $\dfrac{12}{30} \cdot \dfrac{9}{8} \cdot \dfrac{4}{15}$

Solution $\dfrac{\overset{3}{\cancel{12}}}{\underset{\underset{5}{\cancel{10}}}{\cancel{30}}} \cdot \dfrac{\overset{\overset{1}{\cancel{3}}}{\cancel{9}}}{\underset{\underset{1}{\cancel{2}}}{\cancel{8}}} \cdot \dfrac{\overset{\overset{1}{\cancel{2}}}{\cancel{4}}}{\underset{5}{\cancel{15}}} = \dfrac{3 \cdot 1 \cdot 1}{5 \cdot 1 \cdot 5} = \underline{\underline{\dfrac{3}{25}}}$ **STEP 1:** Look for all cancellations.

STEP 2: Multiply.

Notice this time we did *all* the cancellation in the same step. This is how we expect you to do it. However, for the purpose of showing you how cancellation is done, its easier to show *one* set of cancellations at a time. Therefore, the above problem was done this way:

$\dfrac{\overset{3}{\cancel{12}}}{30} \cdot \dfrac{9}{\underset{2}{\cancel{8}}} \cdot \dfrac{4}{15}$

then: $\dfrac{\overset{3}{\cancel{12}}}{\underset{10}{\cancel{30}}} \cdot \dfrac{\overset{3}{\cancel{9}}}{\underset{2}{\cancel{8}}} \cdot \dfrac{4}{15}$

then: $\dfrac{\overset{3}{\cancel{12}}}{\underset{\underset{1}{\cancel{10}}}{\cancel{30}}} \cdot \dfrac{\overset{3}{\cancel{9}}}{\underset{1}{\cancel{8}}} \cdot \dfrac{\overset{2}{\cancel{4}}}{15}$

then: $\dfrac{\overset{3}{\cancel{12}}}{\underset{5}{\underset{\cancel{10}}{\cancel{30}}}} \cdot \dfrac{\overset{3}{\cancel{9}}}{\underset{1}{\cancel{8}}} \cdot \dfrac{\overset{\overset{1}{\cancel{2}}}{\cancel{4}}}{15}$

then: $\dfrac{\overset{3}{\cancel{12}}}{\underset{5}{\underset{\cancel{10}}{\cancel{30}}}} \cdot \dfrac{\overset{\overset{1}{\cancel{3}}}{\cancel{9}}}{\underset{1}{\underset{\cancel{2}}{\cancel{8}}}} \cdot \dfrac{\overset{\overset{1}{\cancel{2}}}{\cancel{4}}}{\underset{5}{\cancel{15}}} = \dfrac{3 \cdot 1 \cdot 1}{5 \cdot 1 \cdot 5} = \underline{\underline{\dfrac{3}{25}}}$

Knowing the Tests for Divisibility (3-2) comes in very handy when cancelling.

EXERCISE 5-1

Cancel whenever possible. Use Tests for Divisibility to help (3-2).

*1. $\dfrac{3}{5} \cdot \dfrac{4}{3} =$

2. $\dfrac{6}{13} \cdot \dfrac{7}{9} =$

3. $\dfrac{21}{25} \cdot \dfrac{8}{11} =$

*4. $\dfrac{15}{16} \cdot \dfrac{48}{65} =$

5. $\dfrac{3}{10} \cdot \dfrac{5}{7} =$

6. $\dfrac{36}{37} \cdot \dfrac{1}{8} =$

7. $\dfrac{4}{9} \cdot \dfrac{5}{9} =$

*8. $\dfrac{16}{23} \cdot \dfrac{16}{23} =$

9. $\dfrac{17}{20} \cdot \dfrac{100}{134} =$

*10. $\dfrac{37}{50} \cdot \dfrac{5}{74} =$

*11. $\dfrac{3}{8} \cdot \dfrac{5}{4} \cdot \dfrac{32}{15} =$

*12. $\dfrac{15}{16} \cdot \dfrac{8}{3} \cdot \dfrac{4}{25} =$

13. $\dfrac{60}{63} \cdot \dfrac{9}{10} \cdot \dfrac{49}{80} =$

14. $\dfrac{2}{9} \cdot \dfrac{2}{3} \cdot \dfrac{4}{7} =$

15. $\dfrac{6}{7} \cdot \dfrac{42}{48} \cdot \dfrac{16}{24} =$

*16. $\dfrac{1000}{5555} \cdot \dfrac{5}{10000} \cdot \dfrac{2}{3} =$

*17. $\dfrac{45}{63} \cdot \dfrac{5}{6} \cdot \dfrac{30}{105} =$

*18. $\dfrac{3}{7} \cdot \dfrac{4}{9} \cdot \dfrac{5}{8} \cdot \dfrac{7}{16} =$

*19. $\dfrac{2}{15} \cdot \dfrac{45}{49} \cdot \dfrac{28}{30} \cdot \dfrac{1}{2} =$

*20. $\dfrac{2}{3} \cdot \dfrac{2}{3} \cdot \dfrac{2}{3} \cdot \dfrac{3}{2} =$

Don't forget to check your answers before continuing on!

IMPORTANT: Cancellation is a very important tool. You not only use it now, but you will also use it in Algebra—so learn it well. Also, don't forget to use the *Tests for Divisibility* (3-2) to help with the cancellation process.

FRACTIONS II: MULTIPLICATION AND DIVISION   155

SECTION 5-2 MULTIPLICATION OF MIXED NUMBERS

Now that you know how to multiply fractions, let's find out how to multiply mixed number fractions.

GIVEN $2\frac{1}{3} \cdot 5\frac{3}{7}$

Solution   **STEP 1**   Change mixed numbers into improper fractions. (See 4-3)

$$2\frac{1}{3} \cdot 5\frac{3}{7} = \frac{7}{3} \cdot \frac{38}{7}$$

STEP 2   Look for cancellation.

$$2\frac{1}{3} \cdot 5\frac{3}{7} = \frac{\cancel{7}^{1}}{3} \cdot \frac{38}{\cancel{7}_{1}}$$

STEP 3   Multiply numerators together, then multiply denominators together.

$$2\frac{1}{3} \cdot 5\frac{3}{7} = \frac{\cancel{7}^{1}}{3} \cdot \frac{38}{\cancel{7}_{1}} = \frac{1 \cdot 38}{3 \cdot 1} = \frac{38}{3} = 12\frac{2}{3}$$

GIVEN $4\frac{3}{4} \cdot 1\frac{8}{9}$

Solution $4\frac{3}{4} \cdot 1\frac{8}{9} = \frac{19}{4} \cdot \frac{17}{9}$ **STEP 1:** Change mixed numbers into improper fractions.

$4\frac{3}{4} \cdot 1\frac{8}{9} = \frac{19}{4} \cdot \frac{17}{9} = \frac{19 \cdot 17}{4 \cdot 9} = \frac{323}{36}$ **STEPS 2 AND 3:** Look for cancellation, then multiply.

$\therefore 4\frac{3}{4} \cdot 1\frac{8}{9} = \frac{323}{36} = 8\frac{35}{36}$

GIVEN $16\frac{2}{3} \cdot 8\frac{4}{5} \cdot \frac{7}{8}$

Solution $16\frac{2}{3} \cdot 8\frac{4}{5} \cdot \frac{7}{8} = \frac{50}{3} \cdot \frac{44}{5} \cdot \frac{7}{8}$ **STEP 1:** Change mixed numbers into improper fractions.

$16\frac{2}{3} \cdot 8\frac{4}{5} \cdot \frac{7}{8} = \frac{\cancel{50}^{5}}{3} \cdot \frac{\cancel{44}^{11}}{\cancel{5}_{1}} \cdot \frac{7}{\cancel{8}_{2}} = \frac{5 \cdot 11 \cdot 7}{3 \cdot 1 \cdot 1} = \frac{385}{3}$ **STEPS 2 AND 3:** Look for cancellation, then multiply.

$$\therefore 16\frac{2}{3} \cdot 8\frac{4}{5} \cdot 7\frac{7}{8} = \frac{385}{3} = 128\frac{1}{3}$$

GIVEN $\quad 3 \cdot 4\frac{5}{8} \cdot \frac{7}{9}$

Solution $\quad 3 \cdot 4\frac{5}{8} \cdot \frac{7}{9} = \frac{\overset{1}{\cancel{3}}}{1} \cdot \frac{37}{8} \cdot \frac{7}{\underset{3}{\cancel{9}}} = \frac{1 \cdot 37 \cdot 7}{1 \cdot 8 \cdot 3} = \frac{259}{24} = 10\frac{19}{24}$

EXERCISE 5-2

*1. $2\frac{1}{3} \cdot 4\frac{5}{8} =$

2. $3\frac{5}{9} \cdot 1\frac{2}{5} =$

3. $6\frac{7}{11} \cdot 7\frac{7}{10} =$

*4. $5\frac{1}{2} \cdot \frac{3}{4} =$

5. $12\frac{1}{4} \cdot 2\frac{7}{9} =$

*6. $14\frac{2}{7} \cdot 1\frac{19}{30} \cdot 3 =$

7. $5 \cdot 2\frac{1}{7} \cdot 3\frac{6}{15} =$

8. $7\frac{7}{9} \cdot 3\frac{6}{7} \cdot \frac{1}{3} =$

9. $5\frac{1}{3} \cdot 8\frac{1}{4} \cdot 1\frac{1}{22} =$

*10. $25\frac{1}{2} \cdot \frac{10}{17} \cdot 1\frac{2}{45} =$

11. Find the product of $12\frac{3}{8}$ and $2\frac{2}{3}$.

12. Find the product of $7\frac{1}{9}$ and $3\frac{3}{4}$.

13. Find the product of 6, $17\frac{1}{10}$ and $\frac{5}{6}$.

14. Find the product of $1\frac{3}{40}$, $\frac{8}{9}$ and $3\frac{2}{9}$.

*15. Find the product of $18\frac{2}{5}$, $2\frac{5}{11}$ and $3\frac{3}{10}$.

*16. On a blueprint $1\frac{1}{4}$ inch represents 1 foot. How long a line must be drawn to represent $4\frac{5}{8}$ feet?

*17. A coat that originally cost $25 is now marked $\frac{1}{5}$ off. How much money will you save if you buy the coat now?

SECTION 5-3 DIVISION OF FRACTIONS

GIVEN $\dfrac{5}{9} \div \dfrac{3}{5}$

Solution

STEP 1 Invert the fraction that "follows" (is to the right of) the division sign and change the operation to multiplication.

$$\dfrac{5}{9} \div \dfrac{3}{5} = \dfrac{5}{9} \cdot \dfrac{5}{3}$$

STEP 2 Do all possible cancellations. (There are none in this problem.)

STEP 3 Multiply numerators together, then multiply denominators together.

$$\dfrac{5}{9} \div \dfrac{3}{5} = \dfrac{5}{9} \cdot \dfrac{5}{3} = \dfrac{25}{27}$$

Division of fractions is changed into multiplication of fractions by inverting ("flipping over") any fraction that immediately follows a division sign. Once the inversion takes place, the operation is no longer division. It is now multiplication and the problem is subject to cancellation rules.

IMPORTANT!! *You cannot cancel in a division problem.* You must first change the problem into multiplication by inverting any fraction following a division sign ($\div$) and changing the division sign to multiplication ($\bullet$).

GIVEN $\dfrac{6}{7} \div \dfrac{18}{21}$

Solution

$\dfrac{6}{7} \div \dfrac{18}{21} = \dfrac{6}{7} \cdot \dfrac{21}{18}$

STEP 1: Invert fraction following division sign and change operation to multiplication.

$\dfrac{6}{7} \div \dfrac{18}{21} = \dfrac{\cancel{6}^1}{\cancel{7}_1} \cdot \dfrac{\cancel{21}^1}{\cancel{18}_3} = \dfrac{1 \cdot 1}{1 \cdot 1} = 1$

STEPS 2 AND 3: Cancel and proceed as for multiplication of fractions problem.

$\therefore \dfrac{6}{7} \div \dfrac{18}{21} = 1$

GIVEN $\dfrac{7}{8} \div \dfrac{4}{49} \div \dfrac{21}{26}$

Solution $\dfrac{7}{8} \div \dfrac{4}{49} \div \dfrac{21}{26} = \dfrac{7}{8} \cdot \dfrac{\overset{*}{49}}{4} \cdot \dfrac{\overset{*}{26}}{21}$

STEP 1: Invert *any* fraction immediately following a *division* sign and change operation to multiplication.

* Both of these fractions get inverted.

$$\dfrac{7}{8} \div \dfrac{4}{49} \div \dfrac{21}{26} = \dfrac{\overset{1}{\cancel{7}}}{8} \cdot \dfrac{49}{\underset{2}{\cancel{4}}} \cdot \dfrac{\overset{13}{\cancel{26}}}{\underset{3}{\cancel{21}}} = \dfrac{49 \cdot 13}{8 \cdot 2 \cdot 3} = \dfrac{637}{48} = 13\dfrac{13}{48}$$

STEPS 2 AND 3: Cancel, then complete the problem.

$\therefore \dfrac{7}{8} \div \dfrac{4}{49} \div \dfrac{21}{26} = \underline{\underline{\dfrac{637}{48}}} = \underline{\underline{13\dfrac{13}{48}}}$

GIVEN $\dfrac{9}{10} \div \dfrac{45}{51} \div \dfrac{17}{20}$

Solution $\dfrac{9}{10} \div \dfrac{45}{51} \div \dfrac{17}{20} = \dfrac{9}{10} \cdot \dfrac{\overset{*}{51}}{45} \cdot \dfrac{\overset{*}{20}}{17} = \dfrac{\overset{1}{\cancel{9}}}{\underset{1}{\cancel{10}}} \cdot \dfrac{\overset{3}{\cancel{51}}}{\underset{5}{\cancel{45}}} \cdot \dfrac{\overset{2}{\cancel{20}}}{\underset{1}{\cancel{17}}} = \underline{\underline{\dfrac{6}{5}}} = \underline{\underline{1\dfrac{1}{5}}}$

* Both these fractions get inverted since *both* follow (are to the right of) division signs.

GIVEN $\dfrac{13}{15} \div \dfrac{2}{5} \cdot \dfrac{4}{26}$

$\dfrac{13}{15} \cdot \dfrac{5}{2} \cdot \dfrac{4}{26} = \dfrac{\overset{1}{\cancel{13}}}{\underset{3}{\cancel{15}}} \cdot \dfrac{\overset{1}{\cancel{5}}}{\underset{1}{\cancel{2}}} \cdot \dfrac{\overset{\overset{1}{\cancel{2}}}{\cancel{4}}}{\underset{\underset{1}{\cancel{2}}}{\cancel{26}}} = \underline{\underline{\dfrac{1}{3}}}$

* Only the fraction immediately following the division sign needed to be inverted.

EXERCISE 5-3

*1. $\dfrac{3}{8} \div \dfrac{5}{9} =$

*2. $\dfrac{4}{5} \div \dfrac{3}{20} =$

3. $\dfrac{16}{25} \div \dfrac{5}{8} =$

4. $\dfrac{7}{10} \div \dfrac{49}{50} =$

5. $\dfrac{15}{22} \div \dfrac{27}{33} =$

*6. $\dfrac{13}{15} \div \dfrac{26}{27} \div \dfrac{5}{9} =$

7. $\dfrac{2}{5} \div \dfrac{36}{45} \div \dfrac{9}{10} =$

8. $\dfrac{7}{10} \div \dfrac{20}{21} \div \dfrac{49}{50} =$

*9. $\dfrac{31}{40} \div \dfrac{5}{62} \bullet \dfrac{2}{25} =$

*10. $\dfrac{3}{4} \div \dfrac{16}{19} \div \dfrac{38}{40} =$

SECTION 5-4 DIVISION OF MIXED NUMBER FRACTIONS

GIVEN $2\frac{1}{3} \div 4\frac{5}{9}$

Solution

STEP 1 Change all mixed numbers into improper fractions. (See 4-3)

$$2\frac{1}{3} \div 4\frac{5}{9} = \frac{7}{3} \div \frac{41}{9}$$

STEP 2 Invert any fraction immediately following division sign and change operation to multiplication.

$$2\frac{1}{3} \div 4\frac{5}{9} = \frac{7}{3} \div \frac{41}{9} = \frac{7}{3} \cdot \frac{9}{41}$$

STEP 3 Complete as before.

$$2\frac{1}{3} \div 4\frac{5}{9} = \frac{7}{3} \div \frac{41}{9} = \frac{7}{\cancel{3}_1} \cdot \frac{\cancel{9}^3}{41} = \frac{7 \cdot 3}{1 \cdot 41} = \underline{\underline{\frac{21}{41}}}$$

$$\therefore 2\frac{1}{3} \div 4\frac{5}{9} = \underline{\underline{\frac{21}{41}}}$$

GIVEN $8\frac{3}{4} \div 2\frac{3}{16}$

Solution

$$8\frac{3}{4} \div 2\frac{3}{16} = \frac{35}{4} \div \frac{35}{16}$$

STEP 1: Change mixed numbers into improper fractions.

$$8\frac{3}{4} \div 2\frac{3}{16} = \frac{35}{4} \div \frac{35}{16} = \frac{35}{4} \cdot \frac{16}{35}$$

STEP 2: Invert and change operation to multiplication.

$$8\frac{3}{4} \div 2\frac{3}{16} = \frac{35}{4} \div \frac{35}{16} = \frac{\cancel{35}^1}{\cancel{4}_1} \cdot \frac{\cancel{16}^4}{\cancel{35}_1} = \frac{1 \cdot 4}{1 \cdot 1} = \frac{4}{1} = \underline{\underline{4}}$$

STEP 3: Complete problem.

$$\therefore 8\frac{3}{4} \div 2\frac{3}{16} = \underline{\underline{4}}$$

FRACTIONS II: MULTIPLICATION AND DIVISION **161**

GIVEN $5\frac{3}{5} \div 1\frac{7}{15} \div 8$

Solution $5\frac{3}{5} \div 1\frac{7}{15} \div 8 = \frac{28}{5} \div \frac{22}{15} \div \frac{8}{1}$ **STEP 1**

$5\frac{3}{5} \div 1\frac{7}{15} \div 8 = \frac{28}{5} \div \frac{22}{15} \div \frac{8}{1} = \frac{28}{5} \cdot \frac{15}{22} \cdot \frac{1}{8}$ **STEP 2**

$5\frac{3}{5} \div 1\frac{7}{15} \div 8 = \frac{28}{5} \div \frac{22}{15} \div \frac{8}{1} = \frac{\overset{7}{\overset{14}{\cancel{28}}}}{\underset{1}{\cancel{5}}} \cdot \frac{\overset{3}{\cancel{15}}}{\underset{11}{\cancel{22}}} \cdot \frac{1}{\underset{4}{\cancel{8}}} = \frac{7 \cdot 3 \cdot 1}{1 \cdot 11 \cdot 4} = \underline{\underline{\frac{21}{44}}}$ **STEP 3**

$\therefore 5\frac{3}{5} \div 1\frac{7}{15} \div 8 = \underline{\underline{\frac{21}{44}}}$

GIVEN $16 \div 5\frac{1}{3} \div 3$

Solution $16 \div 5\frac{1}{3} \div 3 = \frac{16}{1} \div \frac{16}{3} \div \frac{3}{1} = \frac{\overset{1}{\cancel{16}}}{1} \cdot \frac{\overset{1}{\cancel{3}}}{\underset{1}{\cancel{16}}} \cdot \frac{1}{\underset{1}{\cancel{3}}} = \frac{1}{1} = \underline{\underline{1}}$

GIVEN $14\frac{1}{2} \div 13 \div 8\frac{2}{7}$

Solution $14\frac{1}{2} \div 13 \div 8\frac{2}{7} = \frac{29}{2} \div \frac{13}{1} \div \frac{58}{7} = \frac{\overset{1}{\cancel{29}}}{2} \cdot \frac{1}{13} \cdot \frac{7}{\underset{2}{\cancel{58}}} = \underline{\underline{\frac{7}{52}}}$

REMEMBER: When you multiply and divide fractions you do *not* need an LCD. You *must*, however, turn *all* mixed or whole numbers into improper fractions *before* you multiply or divide.

EXERCISE 5-4

*1. $3\frac{2}{3} \div 4\frac{5}{9} =$

2. $6\frac{1}{8} \div 3\frac{2}{3} =$

3. $5\frac{2}{5} \div 10\frac{5}{9} =$

4. $1\frac{6}{7} \div 2\frac{3}{14} =$

5. $9\frac{1}{2} \div 5\frac{3}{8} =$

*6. $7 \div 2\frac{1}{7} =$

*7. $8\frac{11}{12} \div 107 =$

8. $10\frac{1}{10} \div 12\frac{1}{5} =$

9. $16 \div \frac{4}{5} =$

10. $2\frac{3}{4} \div 8\frac{1}{16} =$

*11. $3\frac{1}{3} \div 2\frac{5}{9} \div \frac{3}{4} =$

12. $6\frac{7}{8} \div 3\frac{5}{9} \div 1\frac{1}{16} =$

13. $7\frac{1}{4} \div 1\frac{1}{10} \div \frac{5}{8} =$

*14. $16 \div 2\frac{1}{2} \div \frac{3}{8} =$

15. $7\frac{4}{5} \div 2 \div 2\frac{3}{5} =$

*16. Find the quotient of $8\frac{4}{11}$ and $6\frac{1}{4}$.

17. Find the quotient of $2\frac{12}{13}$ and 7.

18. Find the quotient of $4\frac{5}{9}$ and $1\frac{2}{3}$.

19. Find the quotient of 20 and $5\frac{5}{7}$.

20. Find the quotient of $13\frac{1}{2}$ and $4\frac{1}{2}$.

21. On a scale drawing, a distance of 34 miles is represented by a line $2\frac{1}{4}$ inches long. How long should a line be to represent a distance of $148\frac{2}{3}$ miles?

*22. How many 4 inch wide ceramic tiles are needed to cover a floor $23\frac{1}{4}$ feet wide? (**NOTE:** 4 inches $= \frac{1}{3}$ foot.)

23. A radioactive material gives off $2\frac{1}{3}$ milliroentgens of radiation per hour. A dosimeter measures a total dosage of $18\frac{1}{3}$ milliroentgens. For how many hours was the dosimeter irradiated? (Milliroentgen is a unit of intensity for radioactive emission.)

CHAPTER 5

REVIEW TEST #1

1. Change $3\frac{5}{16}$ into an improper fraction. (see 4-3)

2. Change $\frac{46}{8}$ into a mixed number and reduce if possible. (see 4-3)

Reduce answers in 3–10 if possible.

3. $\frac{7}{9} \times \frac{21}{25} \times \frac{15}{28} = \underline{}$ (see 5-1)

4. $\frac{5}{8} \div \frac{3}{5} = \underline{}$ (see 5-3)

5. The product of $12\frac{6}{7}$ and $3\frac{3}{4}$ is $\underline{}$ (see 5-2)

6. The quotient of $9\frac{1}{12}$ and $7\frac{4}{5}$ is $\underline{}$ (see 5-4)

7. $8\frac{1}{8} \div 5 \div 3\frac{2}{9} = \underline{}$ (see 5-4)

8. $15\frac{1}{5} \times \frac{3}{7} = \underline{}$ (see 5-2)

9. A piece of metal expands $\frac{5}{16}$ inch for each 1°C it is heated. If the metal expanded $2\frac{4}{5}$ inches, how many degrees was the metal heated? (see 5-4)

10. If you worked $32\frac{1}{2}$ hours and received $8\frac{1}{4}$ dollars for every hour you worked, how much did you make? (see 5-2)

CHECK YOUR ✓ ANSWERS If you missed more than two (2) problems, correct your mistakes, study the chapter again, *and take Review Test #2.* If you missed two or fewer problems, ask your instructor for the topic quiz on Fractions II.

CHAPTER 5

REVIEW TEST #2

1. Change $6\frac{8}{9}$ into an improper fraction.

2. Change $\frac{58}{8}$ into a mixed number.

Reduce answers in 3–10 if possible.

3. $\frac{4}{15} \times \frac{35}{6} \times \frac{9}{14} = $ _____?_____

4. $\frac{11}{12} \div \frac{22}{35} = $ _____?_____

5. The product of $75\frac{1}{2}$ and $9\frac{1}{3}$ is _____?_____

6. The quotient of $2\frac{5}{8}$ and $6\frac{3}{7}$ is _____?_____

7. $2\frac{7}{9} \div 8\frac{6}{7} \div 3 = $ _____?_____

8. $12\frac{2}{3} \times \frac{6}{11} = $ _____?_____

9. On a scale drawing, $\frac{1}{2}$ inch represents 25 miles. How long should a line be to represent $230\frac{1}{4}$ miles?

10. If distance traveled is equal to the rate of speed times the time ($D = RT$), how far can you travel at $48\frac{1}{2}$ miles per hour for $3\frac{1}{3}$ hours?

CHECK YOUR ANSWERS If you missed more than two problems, correct your mistakes, study the chapter again, and take Chapter Review Test #3. If you missed two or less than two problems, you should be ready for the topic quiz on Fractions II.

CHAPTER 5

REVIEW TEST #3

1. Change $38\frac{1}{3}$ into an improper fraction.

2. Change $\frac{621}{9}$ into a mixed number.

Reduce answers in 3–10 if possible.

3. $\frac{7}{12} \times \frac{15}{8} \times \frac{32}{35} = \underline{}$

4. $\frac{3}{7} \div \frac{28}{9} = \underline{}$

5. The product of $68\frac{2}{3}$ and $7\frac{1}{5}$ is $\underline{}$

6. The quotient of $14\frac{5}{6}$ and $7\frac{1}{6}$ is $\underline{}$

7. $5\frac{5}{8} \times 3\frac{4}{5} \times 6 = \underline{}$

8. $9\frac{12}{13} \times \frac{11}{12} \ \underline{}$

9. If one resistor draws $\frac{2}{3}$ watt, how many resistors are in a series circuit drawing $36\frac{4}{5}$ watts?

10. The velocity of a body moving at constant speed is equal to the distance covered in a given time divided by the time. What is the velocity of a ball which moves $8\frac{1}{2}$ feet in $1\frac{2}{3}$ second?

CHECK YOUR ✓ ANSWERS If you missed more than two problems, *see your instructor* as soon as possible. Otherwise, you should be ready for the topic quiz on Fractions II.

CHAPTER 6

DECIMALS I: ADDITION, SUBTRACTION AND ROUNDING

INTRODUCTION

In this chapter we are going to review decimal notation, rounding of decimal numbers, changing decimals into fractions and vice versa, and also addition and subtraction of decimal numbers.

SECTION 6-1 PLACE VALUE

What is a decimal? A decimal, also called decimal fraction, is another way of expressing a portion of a whole quantity, just as fractions are a way of showing part of a whole. In decimal fractions, however, we are limited as to the denominators our decimal fractions can have. The denominators of decimal fractions will always be powers of 10:

10;	100;	1000;	10,000;	100,000;	1,000,000;	etc.
(10^1)	(10^2)	(10^3)	(10^4)	(10^5)	(10^6)	

Also, decimal fractions are not written such that we have a numerator and denominator. Decimal fractions are written using decimal notation.

Do you remember the place value chart from Chapter 2? Let's look at it again.

PLACE VALUE CHART—MEMORIZE

← hundred millions | ten millions | millions | hundred thousands | ten thousands | thousands | hundreds | tens | ones (units) | . DECIMAL POINT | tenths | hundredths | thousandths | ten thousandths | hundred thousandths | millionths →

Using this place value chart, let's look at a number in decimal notation: 201.9538.
Starting from the largest place value and working our way down, we have:

2 — hundreds	=	200
0 — tens	=	00
1 — unit	=	1
(decimal point)		.
9 — ten*ths* $\left(\frac{9}{10}\right)$	=	.9
5 — hundred*ths* $\left(\frac{5}{100}\right)$	=	.05
3 — thousand*ths* $\left(\frac{3}{1000}\right)$	=	.003
8 — ten thousand*ths* $\left(\frac{8}{10000}\right)$	=	.0008

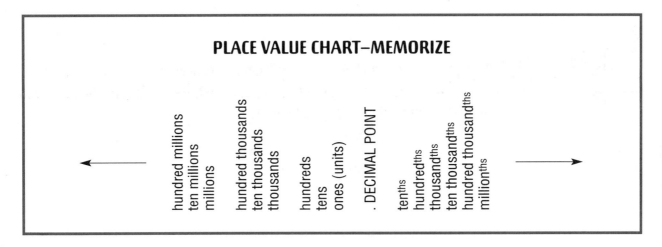

(Another way of showing place value)

hundreds	tens	unit		tenths	hundredths	thousandths	ten thousandths
2	0	1	.	9	5	3	8

DECIMALS I: ADDITION, SUBTRACTION AND ROUNDING 169

Notice that the digits to the right of the decimal point represent *part* of a whole number quantity; whereas, digits to the left of the decimal point represent whole number quantities.

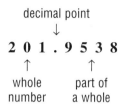

In reading a decimal number, read the whole number part; read "and" to show the placement of the decimal point; read the decimal part as if it were a whole number; then name the place value of the last digit.

EXAMPLE 201.9538 is read "two hundred one *and* nine thousand five hundred thirty-eight *ten-thousandths.*"

"*Ten-thousandths*" in the above example is the place value of the last digit (**8**) in the decimal part of the number.

EXAMPLE .237 is read "two hundred thirty-seven *thousandths.*"

"*Thousandths*" is the place value of the last digit (**7**).

EXAMPLE 12,046.9 is read "twelve thousand forty-six and nine *tenths.*"

EXAMPLE 8.10032 is read "eight and ten thousand thirty-two *hundred thousandths.*"

EXAMPLE .71 is read "seventy-one *hundredths.*"

EXERCISE 6-1

Read the following numbers (that have been written in decimal notation).

1. 895.6
2. .093
3. 2.54
4. 10.03
5. 168.90

6. 36.2835
7. .7005
8. 2,160.178
9. 49.003
10. 1,000,000.000001

170 CHAPTER 6

In problems 11–20, change the words into decimal notation. Example:

GIVEN   Seven hundred six and twenty-three ten thousandths.

Solution   Write the whole number part as you normally would. Replace the "and" with a decimal point. Determine the number of decimal places in the decimal part of the number and put the last digit you read in that place. If you have to fill in the decimal places, use zeros.

"Seven hundred six and twenty-three *ten-thousandths*" = 706.0023

 "*Ten-thousandths*" means four decimal places. Since "three" is the last digit read, it must appear in the "ten thousandths" place.

11. One hundred and two tenths.

12. Twelve thousand sixty-five and thirteen thousandths.

13. Four and twenty-six hundredths.

14. Eight thousand ninety ten thousandths.

15. Nine hundred eighty-three and seven thousandths.

16. One million three hundred thousand twenty-six and thirteen hundredths.

17. Five and six ten thousandths.

18. Seven hundred sixty-two ten thousandths.

19. Thirty-six thousand and thirty-six thousandths.

20. One and one ten thousandth.

SECTION 6-2 DECIMALS TO FRACTIONS AND BACK AGAIN

Let's look at the relationship of fractions and decimal fractions.

Decimal Fraction	Fraction
.8 (read "eight tenths")	$= \dfrac{8}{10}$
.05 (read "five hundredths")	$= \dfrac{5}{100}$
.236 (read "two hundred thirty-six thousandths")	$= \dfrac{236}{1000}$
.9607 (read "nine thousand six hundred seven ten thousandths")	$= \dfrac{9607}{10,000}$
2.4 (read "two and four tenths")	$= 2\dfrac{4}{10}$

As you can see from the examples above, changing a decimal into a fraction is fairly easy. All that needs to be done is to read the decimal.

GIVEN Change .35 into a fraction.

Solution .35 is read "thirty-five hundredths"

$$\therefore 35 = \dfrac{35}{100}$$

$\dfrac{35}{100}$ can be reduced $\quad \therefore \dfrac{35}{100} = \underline{\underline{\dfrac{7}{20}}}$

GIVEN Change .0482 into a fraction.

Solution $.0482 = \dfrac{482}{10,000} = \underline{\dfrac{241}{5000}}$

GIVEN Change 3.706 into a fraction.

Solution $3.706 = 3\dfrac{706}{1000} = 3\underline{\dfrac{353}{500}}$

Now how do you change a fraction into a decimal fraction? Although the procedure you are about to learn is not as easy as changing a decimal into a fraction, it is not hard as long as you remember this: *A fraction is another way of expressing division.*

$$\frac{3}{4} = 3 \div 4 = 4\overline{)3}$$

↑ This is what we are going to do.

▣ Procedure for Changing a Fraction into a Decimal

STEP 1
Write the fraction as a whole number division.

$$\frac{3}{4} = 4\overline{)3}$$

STEP 2
Add a decimal point and some zeros. (You can always add more zeros if needed.)

$$4\overline{)3.00}$$

STEP 3
Place the decimal point directly above in your answer and proceed with the division.

$$\begin{array}{r} .75 \\ 4\overline{)3.00} \\ \underline{2\ 8} \\ 20 \\ \underline{20} \end{array}$$

∴ $\frac{3}{4} = .75$ (This is called a terminating decimal fraction.)

GIVEN   Change $\frac{2}{5}$ into a decimal.

Solution $\frac{2}{5} = 5\overline{)2}$ **STEP 1:** Write as whole number division.

$$5\overline{)2.00}$$ **STEP 2:** Add decimal point and some zeros.

$$\begin{array}{r} .4 \\ 5\overline{)2.00} \\ \underline{2\ 0} \end{array}$$ **STEP 3:** Place decimal point directly above and divide.

∴ $\frac{2}{5} = .4$ (This is another example of a terminating decimal.)

GIVEN   Change $\frac{8}{9}$ into a decimal.

Solution $\frac{8}{9} = 9\overline{)8}$ **STEP 1**

DECIMALS I: ADDITION, SUBTRACTION AND ROUNDING   **173**

$$\begin{array}{r} .888 \\ 9\overline{)8.000} \\ \underline{7\,2} \\ 80* \\ \underline{72} \\ 80* \\ \underline{72} \\ 8* \end{array}$$

STEPS 2 AND 3

* The remainder is going to continue to be the same ∴ the last digit in our answer is going to continue to be the same—or *repeat*. This decimal number is called a repeating decimal or non-terminating decimal. To show that our answer repeats, place a bar above the number (or series of numbers) that repeats.

∴ $\frac{8}{9}$ = .88$\overline{8}$ (This is a repeating decimal fraction.)

GIVEN   Change $\frac{2}{3}$ into a decimal.

Solution $\frac{2}{3} = 3\overline{)2.000}$
$$\begin{array}{r} .666 \\ \underline{1\,8} \\ 20* \\ \underline{18} \\ 20* \\ \underline{18} \\ 2* \end{array}$$

STEPS 1, 2, AND 3

∴ $\frac{2}{3}$ = .66$\overline{6}$ (This is another example of a repeating decimal.)

GIVEN   Change $\frac{7}{18}$ into a decimal.

Solution $\frac{7}{18} = 18\overline{)7.000}$
$$\begin{array}{r} .388 \\ \underline{5\,4} \\ 160* \\ \underline{144} \\ 160* \\ \underline{144} \\ 16* \end{array}$$

Remainder keeps repeating ∴ the last digit in our answer is repeating.

∴ $\frac{7}{18}$ = .38$\overline{8}$ (Another example of a repeating decimal.)

174 CHAPTER 6

GIVEN   Change $\frac{3}{7}$ into a decimal.

Solution

$$\frac{3}{7} = 7\overline{)3.0000}^{.4285}$$

```
       .4285
    7)3.0000
      2 8
        20
        14
         60
         56
          40
          35
           5
```

STEPS 1, 2, AND 3

← There is no evidence yet of the remainder repeating or the division ending ∴ we will assume this is a non-terminating decimal with no sign of repeat. Treatment of this type of decimal will be discussed later in this chapter. For now, simply take the division out to *four decimal places* (ten thousandths) and place three periods after the last digit to show that the decimal continues on.

∴ $\frac{3}{7}$ = .4285 . . .

GIVEN   Change $\frac{6}{11}$ into a decimal.

Solution

```
          .5454
$\frac{6}{11}$ = 11)6.0000
         5 5
          50*
          44
           60**
           55
            50*
            44
             6**
```

STEPS 1, 2, AND 3

A pattern is forming. The **54** is being repeated in the answer ∴ place the bar over the $\overline{54}$ to show that this series of numbers is being repeated.

∴ $\frac{6}{11}$ = .54$\overline{54}$ (This is another repeating or non-terminating decimal.)

EXERCISE 6-2

In problems 1–20 change the given decimal into a fraction.

Example: $40.5 = 40\frac{5}{10} = 40\frac{1}{2}$

1. .5
2. 3.45
3. .09
4. 16.125
5. .1536
6. 10.083
7. .9999
8. 3,654.28367
9. .7003
10. .0005
11. 2.8125
12. .25
13. 19.5
14. .00808
15. 100.001
16. 255.75
17. .875
18. .96875
19. 20.4375
20. 144.2

In problems 21–40, change the given fraction into a decimal.

Example:
$$\frac{5}{9} = 9\overline{)5.000} = .55\overline{5}$$
$$\begin{array}{r} .555 \\ \underline{4\ 5} \\ 50 \\ \underline{45} \\ 50 \\ \underline{45} \\ 5 \end{array}$$

*21. $\frac{3}{8}$

22. $\frac{1}{4}$

*23. $\frac{7}{9}$

24. $\frac{2}{3}$

25. $\frac{2}{5}$

26. $\frac{1}{10}$

27. $\frac{3}{4}$

28. $\frac{1}{2}$

*29. $\frac{13}{16}$

30. $\frac{4}{5}$

*31. $\frac{3}{22}$

*32. $\frac{5}{7}$

33. $\frac{4}{5}$

34. $\frac{1}{8}$

35. $\frac{1}{16}$

*36. $\frac{1}{9}$

37. $\frac{3}{11}$

38. $\frac{7}{100}$

*39. $\frac{23}{1000}$

*40. $\frac{13}{14}$

SECTION 6-3 ROUNDING

Procedure for Rounding a Decimal to a Given Place Value

GIVEN Round .8326 to the nearest thousandth.

Solution

STEP 1 Locate the digit in the given place value:

.8326
↑ 2 is in the thousandths place.

STEP 2 Look at the number directly to the *right* of the place value you're rounding to:

*
.8326
↑ 6 is the number to the right of the thousandths place.

If this number is equal to or greater than 5 (# ≥ 5), round the digit in the place value up to the next digit; and drop all digits to the right of the given place value. Since 6 is ≥ 5 we round up the 2.

*
.8326
↑ 2 must be rounded up to 3.

∴ .8326 = .833 when rounded to the nearest thousandth.

GIVEN Round .8326 to the nearest hundredth.

Solution

.8326
↑ 3 is in the hundredths place.

STEP 1: Locate digit in the given place value.

*
.8326
↑

STEP 2: 2 is the number directly to the right of our given place value. Since 2 is not greater than or equal to 5, we do *not* round the 3 up. We leave it alone and drop the digits to the right of the 3.

∴ .8326 = .83 when rounded to the nearest hundredth.

GIVEN Round 12.0862 to the nearest thousandth; to the nearest hundredth; to the nearest tenth.

Solution

12.0862
↑

STEP 1: Rounding to the nearest thousandth makes 6 our digit to round.

```
         *
   12.0862                    STEP 2:  2 is the number to the right of 6, and since 2 is not
       ↑                               equal to or greater than 5, we leave the 6 alone and
                                       simply drop the digit(s) to the right of the 6.
```

∴ 12.0862 = <u>12.086</u> when rounded to the nearest thousandth.

To the nearest hundredth?

```
   12.0862                    STEP 1:  8 is in the hundredths place.
       ↑

        *
   12.0862                    STEP 2:  6 is the number we look at. Since 6 *is* greater than 5,
       ↑                               we must round our 8 up to 9 and drop the digits to
                                       the right of the hundredths place.
```

∴ 12.0862 = <u>12.09</u> when rounded to the nearest hundredth.

To the nearest tenth?

```
   12.0862                    STEP 1:  0 is in the tenths place.
     ↑

      *
   12.0862                    STEP 2:  8 is the number we look at. Since 8 is greater than 5,
     ↑                                 we must round the 0 up to 1 and drop the digits that
                                       follow.
```

∴ 12.0862 = <u>12.1</u> when rounded to the nearest tenth.

Round 100.9536 to:

A. Nearest whole number.
B. Nearest tenth.
C. Nearest hundredth.
D. Nearest thousandth.

Solution

A. To the nearest whole number:

```
   100.9536                   STEP 1:  0 is in the whole number place.
      ↑

       *
   100.9536                   STEP 2:  9 is in the place directly to the right of where we are
      ↑                                rounding. Since 9 is greater than 5, we must round the
                                       0 up to 1 and drop all digits to the right of the whole
                                       number place.
```

∴ 100.9536 = <u>101</u> when rounded to the nearest whole number.

B. To the nearest tenth:

100.9536
↑
STEP 1: 9 is in the tenths place.

*
100.9536
↑
STEP 2: 5 is the number we look at. Since 5 is equal to 5, we round the 9 up to 10.

∴ 100.9536 = 101.0 when rounded to the nearest tenth.

C. To the nearest hundredth:

100.9536
↑
STEP 1: 5 is in the hundredths place.

*
100.9536
↑
STEP 2: 3 is the number we look at. Since 3 is less than 5 (3 < 5), we leave the digit we are rounding alone.

∴ 100.9536 = 100.95 when rounded to the nearest hundredth.

D. To nearest thousandth:

100.9536
↑
STEP 1: 3 is in the thousandths place.

*
100.9536
STEP 2: 6 is the number we look at. Since 6 is greater than 5 (6 > 5), we round the 3 up to 4 and drop the digits to the right of the thousandths place.

∴ 100.9536 = 100.954 when rounded to the nearest thousandth.

GIVEN Round 99.99995 to the nearest ten-thousandth

Solution

*
99.99995
↑

Add a one to this place
1 ←
99.9990
↑ Becomes a zero

1 ←
99.9900
↑ Becomes a zero

STEP 1 AND 2: 9 is in the ten thousandths place. 5 is the number we look at. Since 5 = 5, we round up the 9 to 10, by placing a zero (0) in the ten thousandths place and adding a one (1) to the number in the thousandths place. Also, at this point we drop the digit (5) to the right of the ten thousandths place.

Since the digit in the thousandths place is a 9, adding a 1 to it causes this 9 to be replaced by a 0 and a 1 to be carried over to the hundredths place.

This continues on down the line, until finally our answer becomes: 100.0000

∴ 99.99995 rounded to the nearest ten thousandth is 100.0000

NOTE: When no other instructions are given, the standard rule of rounding is to round to the nearest thousandth. An exception to this "rule of thumb" is when working with money. Money is usually rounded to the nearest hundredth.

EXERCISE 6-3

Round all of the following numbers to:
a. *Nearest whole number.*
b. *Nearest tenth.*
c. *Nearest hundredth.*
d. *Nearest thousandth.*

1. 6.7325
2. 18.0932
3. 20.1234
4. 4.3549
5. 22.5555
6. 129.9999
7. 1.5678
8. 48.2103
9. 2873.4568
10. 77.4709
11. 456.9997
12. 210.0095
13. 3400.4672
14. 986.6890
15. 3.1615
16. 80.5712
17. 79.9832
18. 6.3261
19. 9.5484
20. 1000.1001

SECTION 6-4 ADDITION AND SUBTRACTION OF DECIMALS

▣ Procedure for Adding Decimals

GIVEN   12.83 + 7.9 + .682

Solution

STEP 1   Rewrite the given problem such that the decimal points in every number line up.

```
   12.83      ← (two decimal places)
    7.9       ← (one decimal place)
+    .682     ← (three decimal places)
```
↑ Decimal points must line up!

STEP 2   Bring the decimal point down in the answer and add the digits as in whole numbers.

```
   12.83
    7.9
+    .682
   21.412
```
↑ Decimal points all line up.

NOTE: If it bothers you that these decimals do not have the same number of decimal places, add zeros as place holders so that the decimals do all have the same number of places.

```
   12.830     ← (three decimal places)
    7.900     ← (three decimal places)
+    .682     ← (three decimal places)
   21.412
```

GIVEN   .0593 + 200 + 4.65

Solution

```
      .0593
  *200.
+    4.65
```

STEP 1: Line up decimal points. *200 is a whole number. All whole numbers can be written with a decimal point at the end of the number (to the right of the last digit).

```
      .0593
  *200.0000
+    4.6500
   204.7093
```

STEP 2: Bring down decimal point and add digits. (Add zeros to decimals if desired.)

DECIMALS I: ADDITION, SUBTRACTION AND ROUNDING

GIVEN 1038.62 + 2564.9 + 14.8523

Solution

```
   1038.62
   2564.9
+    14.8523
   3618.3723
```

STEPS 1 AND 2: Line up decimal points. Bring decimal point down and add. Add zeros if desired.

GIVEN The following weights were added to a balance in order to bring it into balance: 7.0001 g, 3.1204 g, 5.4300 g, .4030 g. What is the total weight added to the balance? (g = gram, a metric unit of weight.)

Solution Since the weights were *added* to the balance and we are asked to find the total of these weights, we are going to add the given weights together.

```
    7.0001 g
    3.1204 g
    5.4300 g
+    .4030 g
   15.9535 g
```

∴ The total weight added to the balance is 15.9535 g.

▣ Procedure for Subtracting Decimals

GIVEN 15.53 − 12.9

Solution

STEP 1 As in addition of decimals, we must line up the decimal points.

```
   15.53   ← (two decimal places)
−  12.9    ← (one decimal place)
```

STEP 2 Add zeros to the decimal numbers so they have the same number of decimal places.

```
   15.53
−  12.90   ← (We added this zero such that we now have two decimal places in
              both of the numbers.)
```

STEP 3 Bring the decimal point down in the answer and subtract the digits as in whole numbers.

```
   15.53
−  12.90
    2.63
```

> **NOTE:** When you add zeros to a decimal number, place the zeros to the right of the last digit in the decimal number. Adding zeros in this way does not change the value of the number, just it's looks.

$$.3 = .30 = .300 = .3000 \ldots$$

$$.3 = \frac{3}{10}$$

$$.30 = \frac{30}{100} = \frac{3}{10}$$

$$.300 = \frac{300}{1000} = \frac{30}{100} = \frac{3}{10}$$

$$.3000 = \frac{3000}{10,000} = \frac{300}{1000} = \frac{30}{100} = \frac{3}{10}$$

GIVEN $6.5 - 2.385$

Solution

$\quad 6.5$
$-\,2.385$

STEP 1: Line up decimal points.

$\quad 6.500$
$-\,2.385$

STEP 2: Add zeros if needed.

$\quad 6.500$
$-\,2.385$
$\quad \overline{4.115}$

STEP 3: Bring decimal point down and subtract.

GIVEN Find the difference of 35 and 14.69.

Solution

$\quad 35.$
$-\,14.69$

STEP 1: Line up decimal points. (35 = 35.)

$\quad 35.00$
$-\,14.69$

STEP 2: Add zeros.

$\quad 35.00$
$-\,14.69$
$\quad \overline{20.31}$

STEP 3: Bring down decimal point and subtract.

DECIMALS I: ADDITION, SUBTRACTION AND ROUNDING

GIVEN

A student received a grant for $350. With this grant money, the student made the following purchases: $185.69 for tuition, $84.26 for books, and $23.89 for miscellaneous supplies. How much of the grant money does the student have left?

Solution

In order to determine how much money is left, we should determine the total amount of money spent; so add the purchases together first.

$185.69
$ 84.26
$ 23.89
───────
$293.84 = Total amount of purchases

Now to find how much the student has left, we should subtract this total amount of purchases from the $350.

$350.00
− 293.84
───────
$ 56.16

∴ The student has **$56.16** of the grant remaining after his purchases.

EXERCISE 6-4

Perform the given operation(s).

1. 15.9 + 16.83 + 17.921
*2. 28 + .005 + 6.8
3. .523 + .01 + .637
4. 4.698 + 3.521 + .05
*5. 1.93 + 8 + .2134
*6. 130 + .998 + 3.7
*7. .01023 + .52 + 47
8. 2.6 + 3.051 + .00001
9. 3 + 18.912 + 16.7 + 21.003 + .0008
10. .13594 + 116 + 8.2
11. 21.8 − .7
12. .035 − .004
*13. 13.59 − .123
14. 2.713 − .8
15. 16.92 − 14.1
16. 320.19 − 6.23
*17. 25 − 3.84
*18. 4.0538 − 2
19. 78.051 − .6
20. 113.9234 − 111.375

*21. Find the sum of 7.6, 18.92, 14.7 and .003.
22. Find the sum of 21, 118.1, .659 and 7.
*23. Find the difference of 43 and .21.
24. Find the difference of 2.01 and .7935.
25. Find the difference of 4.359 and 1.

*26. A compound consists of sodium (Na), nitrogen (N), and oxygen (O). How much does a mole weigh if it contains the following: 22.99 g Na, 14.007 g N, 48 g O? (A mole is a chemical unit of quantity.)

*27. In a parallel electrical circuit (see below), the sum of the current in each leg equals the total amount of current in the circuit. If the total current is 15.70 amps and the current in two of the legs is 4.35 and 6.94 amps respectively, calculate the current in the third leg.

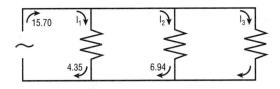

*28. On a timed math quiz, you were able to answer 50 questions in 55.8 minutes. On the same quiz, given a week before, you completed the 50 questions in 62.3 minutes. How much faster were you the second time you took the quiz?

CHAPTER 6

REVIEW TEST #1

1. Write in decimal notation: (see 6-1)

    Fourteen thousand five hundred and eighty-two thousandths.

2. Change .8125 into a fraction. (see 6-2)

3. Change $\frac{5}{8}$ into a decimal. (see 6-2)

4. Round 5.0986 to the nearest: (see 6-3)
    a. Thousandth.
    b. Hundredth.
    c. Tenth.
    d. Whole number.

5. The sum of 12.2, 3.98 and .706 is ___?___ (see 6-4)

6. The difference of 320.9 and 143.08 is ___?___ (see 6-4)

7. 612 + 9.73 = ___?___ (see 6-4)

8. 87 − .56 = ___?___ (see 6-4)

9. 4.8 + 19.73 + 25 + .0536 = ___?___ (see 6-4)

10. 45.863 − .5 = ___?___ (see 6-4)

11. Total resistance in a series circuit is equal to the sum of the individual resistances (see 6-4)
    ($R = R_1 + R_2 + R_3 + \ldots$). If the individual resistances are 3.7, 8.9 and 4.5 ohms, what is the total resistance in the circuit?

CHECK YOUR ANSWERS — If you missed three (3) or more answers, correct your mistakes, study the chapter again, and take Review Test #2. If you missed less than three (3) answers, correct your mistakes and ask your instructor for the quiz on Decimals I.

CHAPTER 6

Review Test #2

1. Write in decimal notation:

    Five thousand sixty-seven and one thousand seventy-three ten thousandths.

2. Change 2.55 into a fraction.

3. Change $\frac{7}{9}$ into a decimal.

4. Round 28.5067 to the nearest:
    a. Thousandth.
    b. Hundredth.
    c. Tenth.
    d. Whole number.

5. The sum of 28.6, 3.751, and .05 is ___?___

6. The difference of 6.952 and 3.41 is ___?___

7. 263 + .751 = ___?___

8. 407 − .83 = ___?___

9. 83.95 + 4.8 + 13.752 + 2 = ___?___

10. 698.4 − .0513 = ___?___

11. Total depreciation on an item is equal to the purchase price minus the salvage value ($D = P - S$). If the purchase price on a new car was $8,069.56 and the salvage is $5,697.08, what is the total depreciation on the car?

CHECK YOUR ANSWERS If you missed three (3) or more answers, correct your mistakes and take Review Test #3. Otherwise, correct your mistakes and ask your instructor for the quiz on Decimals I.

CHAPTER 6

REVIEW TEST #3

1. Write in decimal notation:

    Nine hundred and nine hundred thousandths.

2. Change .92 into a fraction.

3. Change $\frac{5}{16}$ into a decimal.

4. Round .8265 to the nearest:

    a. Thousandth.

    b. Hundredth.

    c. Tenth.

    d. Whole number.

5. The sum of 8.7, 136, .463, and 12.82 is ___?___

6. The difference of 346.75 and 6.25 is ___?___

7. 863 + .9 = ___?___

8. 25 − .05 = ___?___

9. 31 + .563 + 14.8914 + 2.6 = ___?___

10. 164.982 − 73.7 = ___?___

11. In one day, a bank made loans to three individuals in the amounts of $1,500, $875.60, and $308.75. How much money did the bank loan in one day?

CHECK YOUR ✓ ANSWERS If you missed three (3) or more answers, *ask your instructor* for further assistance as soon as possible. Otherwise, you should be ready for the topic quiz or Decimals I.

CHAPTERS 1–6

Cumulative Review #2

Use the **Cumulative Reviews** that are spaced throughout this book to help you review and retain the math skills that you have gained. For each of the eight chapters tested on the **Arithmetic Proficiency Test,** you will be required to correctly answer four out of five questions from each chapter. See how you would do with these first six chapters.

Whole Numbers

1. 269 + 4738 + 26 + 3

2. 7581 − 296

3. 183 × 769

4. 64736 ÷ 68

5. 20570 ÷ 34

Symbols and Definitions

1. Product means to _____ ?   (2-2)

2. Write using exponential notation: 2 • 2 • 3 • 5 • 5 • 5   (2-4)

3. Simplify: $3^4 \cdot 5^2$ (2-4)

4. Simplify using order of operations: $4^2 \div (8 - 6) \cdot 3 - 2(14 - 11)$ (2-5)

5. Is $\sqrt{3}$ a rational or irrational number?   (2-7)

Prime Numbers

1. Write the prime factorization of: (3-3)
   a. 190 _____   b. 840 _____

2. Write all the factors of 144 _____ _____   (3-1)

3. Find the Least Common Multiple (LCM) of: (3-5)
   a. 24, 16, 32 _____   b. 49, 84, 105 _____

Fractions I

1. $\dfrac{4}{9} + \dfrac{3}{8} + \dfrac{21}{32}$ (4-6)

2. $\dfrac{65}{67} - \dfrac{2}{3}$ (4-8)

3. $\dfrac{5}{7} + \dfrac{2}{5} - \dfrac{9}{14}$ (4-8)

4. $4\dfrac{2}{15} + 3\dfrac{4}{5} + 1\dfrac{7}{20}$ (4-7)

5. $8\dfrac{7}{17} - 4\dfrac{1}{2}$ (4-9)

Fractions II

1. $\dfrac{12}{19} = \dfrac{?}{38}$ (4-4)

2. $\dfrac{30}{49} \cdot \dfrac{28}{45} \cdot \dfrac{14}{15}$ (5-1)

3. $\dfrac{8}{41} \div \dfrac{3}{4}$ (5-3)

4. $10\dfrac{4}{7} \cdot 3\dfrac{4}{15}$ (5-2)

5. $6\dfrac{1}{2} \div 3\dfrac{1}{3}$ (5-4)

Decimals I

1. Round 149.6972 to the nearest thousandth. (6-3)

2. 8.6 + 491 + .008 (6-4)

3. 147.35 − 83.9 (6-4)

4. Write $\dfrac{39}{10{,}000}$ as a decimal. (6-2)

5. 496 − 2.371 (6-4)

REMEMBER TO CHECK YOUR ANSWERS WITH APPENDIX A IN THE BACK OF THE BOOK.

CHAPTER 7

DECIMALS II: MULTIPLICATION AND DIVISION

INTRODUCTION

In this chapter, you will become reacquainted with multiplication and division of decimals and shortcuts to use when multiplying or dividing by 10 or powers of 10.

SECTION 7-1 MULTIPLICATION OF DECIMALS

Procedure for Multiplying Decimals

GIVEN .72 × 3.4

Solution

STEP 1 Write the problem as you would a whole number multiplication problem.

$$\begin{array}{r} .72 \\ \times\ 3.4 \end{array}$$

Decimal points do not need to be lined up.

STEP 2 Ignore the decimal points for a moment and multiply the numbers as you would whole numbers. Then count the total number of decimal places in each of the numbers you were multiplying. *Starting from the right end of the answer,* count over the total number of decimal places and place the decimal point.

$$\begin{array}{r} .72 \leftarrow\ 2\ \text{decimal places} \\ \times\ 3.4 \leftarrow\ +1\ \text{decimal place} \\ \hline 288\quad\quad\quad 3\ \text{total decimal places} \\ 216\quad\quad\quad \\ \hline 2.448 \leftarrow\ \text{Move the decimal point 3 places to the left.} \end{array}$$

∴ .72 × 3.4 = **2.448**

GIVEN Find the product of 18.73 and .059

Solution

$$\begin{array}{r} 18.73 \\ \times\ .059 \\ \hline \end{array}$$

STEP 1: Write as if whole number multiplication but with decimal points in proper place.

$$\begin{array}{r} 18.73 \leftarrow\ 2\ \text{decimal places} \\ \times\ .059 \leftarrow\ +3\ \text{decimal places} \\ \hline 16857\quad\quad\quad 5\ \text{total decimal places} \\ 9365\quad\quad\quad \\ 0000\quad\quad\quad \\ \hline 1.10507 \leftarrow\ \text{Count over 5 decimal places to the left and place decimal point.} \end{array}$$

STEP 2: Multiply

∴ 18.73 × .59 = **1.10507**

DECIMALS II: MULTIPLICATION AND DIVISION   **193**

GIVEN .03 × .58 × .672

Solution
```
              .58
           × .03
(1st product) .0174
```

STEPS 1 AND 2: Take the first two decimal numbers, multiply, and place decimal point.

```
(1st product)  .0174
                .672
               0348
              01218
              01044
             .0116928
```

REPEAT STEPS 1 AND 2: Take product from first multiplication and multiply it by the last decimal number in the original problem.

GIVEN 12.22 × .002

Solution
```
    12.22
  × .002
   .02444
     ↑
```

STEPS 1 AND 2

Let's take a look at how this zero came to be here.

$$\begin{array}{r} 12.22 \leftarrow \text{(2 decimal places)} \\ \times\ .002 \leftarrow \text{(+3 decimal places)} \\ \hline \text{(5 total decimal places)} \end{array}$$

We know we will need 5 decimal places in the answer, but when we start counting from right to left we find that we have only 4 digits in our answer ∴ we end up with a blank space to the right of where the decimal point should be.

$$\begin{array}{r} 12.22 \\ \times\ .002 \\ \hline .\ 2444 \\ \uparrow \text{ Blank space} \end{array}$$

REMEMBER: Zeros make excellent place holders.

Fill in the blank space with a zero:

$$\begin{array}{r} 12.22 \\ \times\ .002 \\ \hline .02444 \end{array}$$ ∴ .02444 is the answer.

GIVEN $.0005 \times .7$

Solution
$$\begin{array}{rl} .0005 & \text{(4 decimal places)} \\ \times \quad .7 & \text{(1 decimal place)} \\ \hline .00035 & \text{(5 total decimal places)} \end{array}$$

GIVEN   Voltage in a circuit is given by the formula:

$$V = IR \ (I \text{ times } R)$$

where V is the voltage, I is the current in amps and R is the resistance in ohms. Calculate the voltage in a circuit where $I = 2.354$ amps and $R = 10.45$ ohms.

Solution   Since $V = I \times R$, and we are given the value of I as 2.354 amps and the value of R as 10.45 ohms, we must multiply 2.354 by 10.45 to calculate the voltage.

$$\begin{array}{rl} 2.354 & \text{(3 decimal places)} \\ \times \quad 10.45 & \text{(2 decimal places)} \\ \hline 11770 & \text{(5 total decimal places)} \\ 9416 & \\ 23540 & \\ \hline 24.59930 & \leftarrow \text{5 total decimal places} \end{array}$$

∴ The voltage in the circuit V is equal to 24.5993 volts.

EXERCISE 7-1

*1. $.35 \times .6$
2. $4.73 \times .98$
*3. $10 \times .195$
4. $.8 \times .09$
5. 1.953×8.7

*6. 29.5×18
7. $.3456 \times .001$
8. 19.82×1.4
9. $2.005 \times .1$
*10. $.043 \times .0548$

11. $2.6 \times 8.75 \times 3.91$
12. $.09 \times .53 \times 12.8$
13. $.7 \times .3 \times .973$
14. $.008 \times .0004 \times 9.5$
*15. $100 \times .5 \times .83 \times 2.957$

16. Distance D is given by the formula $D = RT$ or $(D = R \times T)$ where R is the rate of speed and T is the amount of time. If $R = 55$ mph (miles per hour) and $T = 3.25$ hours, calculate the distance.

17. If Roy averages 45.6 mph in his new car, how far will he travel in .75 hours (45 minutes)? (Use the $D = RT$ formula.)

18. If a washer is .1875 cm thick, how thick would six washers be?

SECTION 7-2: SHORTCUT FOR MULTIPLYING BY 10 OR POWERS OF 10

Review for a moment what "powers of" 10 means:

10^1—(read "ten to the first power") = 10
10^2—(read "ten squared") = $10 \times 10 = 100$
10^3—(read "ten cubed") = $10 \times 10 \times 10 = 1,000$
10^4—(read "ten to the fourth power") = $10 \times 10 \times 10 \times 10 = 10,000$
and so on.

Let's take a look at what happens to decimal numbers when you multiply by 10 or powers of 10.

GIVEN $10 \times .9$

Solution
```
   10
 × .9
  9.0
```
Notice that .9 becomes 9. when we multiply by 10.

GIVEN $100 \times .9$

Solution
```
  100
 × .9
 90.0
```
.9 becomes 90. when we multiply by 100.

GIVEN $1,000 \times .9$

Solution
```
 1,000
×   .9
 900.0
```
.9 become 900. when we multiply by 1,000.

∴ $10 \times .9 = 9.$ The decimal point moved one place to the right in the decimal number when we multiplied it by 10.

and $100 \times .9 = 90.$ The decimal point moved two places to the right in the decimal number when we multiplied it by 100.

and $1,000 \times .9 = 900.$ The decimal point moved three places to the right when we multiplied by 1,000.

GIVEN $10 \times .387$

Solution
```
  .387
×   10
 3.870
```
The decimal point moved one place to the right when we multiplied the decimal number by 10.

196 CHAPTER 7

GIVEN $100 \times .387$

Solution
```
   .387
×  100
38.700
```
The decimal point moved two places to the right when we multiplied by 100.

GIVEN $1,000 \times .387$

Solution
```
1,000
 .387
```
387.000   The decimal point moved three places to the right when we multiplied by 1,000.

In general:

Multiplication by 10 moves the decimal point *one* place to the *right*.

Multiplication by 100 moves the decimal point *two* places to the *right*.

Multiplication by 1,000 moves the decimal point *three* places to the *right*.

Whatever power of 10 you are multiplying by is the number of places you move the decimal point to the right.

GIVEN $10,000 \times 3.19625$

Solution

Since $10,000 = 10^4$ and since that means we are multiplying 3.19625 by 10^4 we simply move the decimal point 4 places to the right.

(Easy way of showing what power of ten you have is to count the number of zeros after the 1 $\therefore$ 10,000 has 4 zeros, so $10,000 = 10^4$)

$\therefore 10,000 \times 3.19625 = 10^4 \times 3.19625 = 319625\underset{1\,2\,3\,4}{} = \underline{\underline{31962.5}}$

GIVEN $1,000,000 \times 869.735$

Solution

Since 1,000,000 has six zeros and is therefore equal to 10^6, we move the decimal point 6 places to the right.

$\therefore 1,000,000 \times 869.735 = 10^6 \times 869.735 = 869735\underset{1\,2\,3\,4\,5\,6}{} = \underline{\underline{869735000.}}$

GIVEN $1,000,000,000 \times 12.35678$

Solution $1,000,000,000$ equals 10^9 (nine zeros), so . . .

$1,000,000,000 \times 12.35678 = 10^9 \times 12.35678 = 1235678\underset{1\,2\,3\,4\,5\,6\,7\,8\,9}{} = \underline{\underline{123456780000.}}$

EXERCISE 7-2

Use the shortcut method.

*1. $10^2 \times .83$

2. $100 \times .83$

3. $10^3 \times 21.5$

4. $1,000 \times 21.5$

*5. $10^5 \times .0092$

6. $100,000 \times .0092$

7. $10 \times .953$

*8. $.675 \times 10^2$

*9. $100 \times .675$

*10. $2.1 \times 1,000$

11. 2.1×10^3

12. $1,000,000,000 \times 3245.67876$

*13. $10^8 \times .6$

*14. $10^5 \times 625$

15. $10^6 \times 2.73654$

IMPORTANT!! This shortcut method only works when multiplying by *10* or powers of *10*.

SECTION 7-3 DIVISION OF DECIMALS

Now that multiplication by decimals has been covered, it's time to review decimal division.

GIVEN $26.5 \div 2.5$

Solution **STEP 1** Set up problem as in whole number division.

$$2.5 \overline{)26.5}$$

STEP 2 *If* the divisor (in this case 2.5) has any decimal places, move the decimal point in the divisor to the right as many places as needed to make the divisor a whole number. (In this case, the decimal point moves *one* place to the right—2.5 becomes 25.)

$$2.5 \overline{)26.5}$$

STEP 3 Move the decimal point in the dividend (in this case 26.5) the *same* number of places and the same direction (always right) as you moved the decimal point in the divisor.

$$2.5 \overline{)26.5} \quad \therefore 25. \overline{)265.}$$

STEP 4 Place the decimal point in the quotient directly above the new location in the dividend and divide as in whole numbers, adding zeros after the decimal if needed.

$$2.5 \overline{)26.5} = 25. \overline{)265.} = 25 \overline{)265.0} \begin{array}{r} 10.6 \\ \underline{25} \\ 15 \\ \underline{0} \\ 150 \end{array}$$

$$\therefore 26.5 \div 2.5 = \underline{\underline{10.6}}$$

GIVEN $.4386 \div .34$

Solution $.34 \overline{).4386}$ **STEP 1:** Set up problem.

$.34 \overline{).4386}$ **STEPS 2 AND 3:** Move decimal point in divisor to the right as many times as needed to make the divisor a whole number, and move the decimal point in the dividend the *same* number of places.

DECIMALS II: MULTIPLICATION AND DIVISION

$$.34 \overline{)\,.4\overset{.}{3}86}$$

STEP 4: Place decimal point in quotient above new location and divide.

$$.34 \overline{)\,.43\,86}^{\,1.29}$$
$$\underline{34}$$
$$98$$
$$\underline{68}$$
$$306$$
$$\underline{306}$$

and divide.

∴ .4386 ÷ .34 = <u>1.29</u>

GIVEN 58 ÷ .092

Solution

$.092\overline{)58}$ **STEP 1:** Set up.

$.092\overline{)58.\overset{.}{000}}$ **STEPS 2 AND 3:** Move the decimal point in the divisor and move the decimal point in the dividend the same number of places. (Remember a whole number has a decimal point at the right end of the number ∴ 58 = 58. Also, use zeros as place holders.) Place the decimal point in the quotient directly above the new location in the dividend.

$$.092\overline{)58000.0000}^{\,630.4347}$$
$$\underline{552}$$
$$280$$
$$\underline{276}$$
$$40$$
$$\underline{0}$$
$$400$$
$$\underline{368}$$
$$320$$
$$\underline{276}$$
$$440$$
$$\underline{368}$$
$$720$$
$$\underline{644}$$
$$76$$

STEP 4: Divide.

At this point, when there is no sign of our decimal terminating (ending) or repeating, we should stop dividing and round our answer to the nearest thousandth. ∴ 58 ÷ .092 ≅ <u>630.435</u>

IN GENERAL: When performing a decimal division, if your division shows no sign of terminating or repeating, then carry the division out to four decimal places (ten thousandths) and round off to the third decimal place (thousandths). This is a standard rounding rule.

GIVEN 41 ÷ .33

Solution

```
           124.24     ← 24 is being repeated
      33)4100.00              STEPS 1, 2, 3, AND 4
         33
         ‾‾
         80
         66
         ‾‾
         140
         132
         ‾‾‾
          80
          66
          ‾‾
         140
         132
         ‾‾‾
           8
```

A pattern is forming—the digits 24 are repeating in the answer.

∴ 41 ÷ .33 = 124.$\overline{24}$;   rounded to nearest thousandth, 41 ÷ .33 = 124.242

GIVEN 18.92 ÷ 63

Solution   63)‾18.92‾   **STEP 1:** Set up.

Already →   63)‾18.92‾   **STEPS 2 AND 3:** Since there are no decimal places in the divisor, we put the decimal point in the quotient directly above the original location in the dividend.
a whole   ↑
number   Decimal point
             does not move

```
            .3003
      63)18.9200           STEP 4: Divide.
         18 9
         ‾‾‾
            2
            0
            ‾
           20
            0
           ‾‾
          200
          189
          ‾‾‾
           11
```

Does not appear to be a pattern forming, nor does the division appear to be terminating.

∴ 18.92 ÷ 63 ≅ .300   **REMEMBER:** Take division out to at least four *decimal places* and round off to the third if there is no sign of the decimal terminating.

DECIMALS II: MULTIPLICATION AND DIVISION

GIVEN   3.56 ÷ 100

Solution
```
       .0356
100)3.5600
     30 0
      560
      500
       600
       600
```
STEPS 1, 2, 3, AND 4

∴ 3.56 ÷ 100 = .0356

EXERCISE 7-3

1.   1.7 ÷ .25
2.   .9436 ÷ 1.4
3.   .1515 ÷ .075
*4. 36 ÷ .04
5.   968 ÷ .968
6.   .07802 ÷ 9.4
*7. 5.050 ÷ .25
8.   72.072 ÷ 3.6
*9. 22 ÷ .003
10. 11.11 ÷ 9.9
*11. 8 ÷ .51
12. 36.958 ÷ 36.5
*13. 1798.563 ÷ 10
14. .8695 ÷ 1,000
*15. 634.9 ÷ 12
16. 89.5 ÷ 20
17. 13 ÷ .39
18. .65 ÷ .65
*19. 200.01 ÷ .01
20. 387.52 ÷ .0001

*21. The distance between two floors is 8.5 ft. How many steps are needed if each step is .604 ft. high?

*22. A bottle contains 25 g (grams) of medication. If a patient is required to take 1.25 g of medication each day, how long will the bottle last?

SECTION 7-4 SHORTCUT FOR DIVIDING BY 10 OR POWERS OF 10

What happens to a decimal number when divided by 10?

$$2.9 \div 10 = 10\overline{)2.90}$$
$$\underline{2\,0}$$
$$90$$
$$\underline{90}$$

∴ 2.9 ÷ 10 = .29   Notice the digits in the quotient are the same as the dividend but the decimal point in the quotient has moved one place to the left.

$$36.952 \div 10 = 10\overline{)36.9520}$$

Again the digits in the quotient are the same as the dividend but the decimal point moved one place to the left.

How many places and in what direction do you think the decimal point moves if you divided a decimal by 100?

Try this problem and see if you guessed right.

$$7.926 \div 100$$

If you said that division by 100 should move the decimal point *two* places to the *left,* you're right!!

$$100\overline{)7.92600}$$

7.926 becomes .07926 when divided by **100.**

Without actually doing the division, what should the answer be to this problem?

$$450.09 \div 1{,}000$$

If you said that 450.09 ÷ 1,000 should equal .45009, **Congratulations!**

IN GENERAL: Whatever power of ten you are *dividing* by is the number of places you move the decimal point to the *left*.

17.8 ÷ 10 = 1.78	(moved one place to the left)
.05 ÷ 10^1 = .005	
36.5 ÷ 10^2 = .365	(moved two places to the left)
4.896 ÷ 100 = .04896	
8.9 ÷ 10^3 = .0089	(moved three places to the left)
.5432 ÷ 1,000 = .0005432	
23 ÷ 10^4 = .0023	(moved four places to the left)
759.6 ÷ 10,000 = .07596	

Because we are simply moving the decimal point, we typically do not apply rounding rules to these shortcut division problems.

CHAPTER 7

EXERCISE 7-4

Use the shortcut method.

*1. .3 ÷ 10

2. 7.89 ÷ 10

3. 12.0 ÷ 100

*4. 118.6 ÷ 10^2

*5. .395 ÷ 1,000

6. 456.2 ÷ 10^3

7. .03 ÷ 10,000

8. 25.9136 ÷ 10^4

9. .0069 ÷ 10^2

*10. 5.3 ÷ 10^5

*11. 6850 ÷ 100

*12. 4 ÷ 10^3

13. 38.5 ÷ 10^2

*14. .02 ÷ 100

15. 29.736 ÷ 10^2

16. 8147.9 ÷ 10^6

*17. .135 ÷ 1,000,000

18. 7,129,283,470 ÷ 10^9

19. .0003 ÷ 10

*20. 426 ÷ 10^3

IMPORTANT: This shortcut method only works when dividing by 10 or powers of 10.

CHAPTER 7

Review Test #1

1. 3.7 × .5 = __?__ (see 7-1)

2. .6351 ÷ .03 = __?__ (see 7-3)

3. The product of 186 and 3.52 is __?__ (see 7-1)

4. The quotient of 7.032 and 30 is __?__ (see 7-3)

5. 28.6 × 1.982 × .0051 = __?__ (see 7-1)

6. 263 ÷ .263 = __?__ (see 7-3)

Use shortcut method for Problems 7–9.

7. 46.5 × 100 = __?__ (see 7-2)

8. 2 ÷ 100 = __?__ (see 7-4)

9. 65.3786 × 10,000 = __?__ (see 7-2)

10. Unit cost is equal to total cost divided by the number of units. If the total cost of 12 books is $163.08, what is the unit cost? (see 7-3)

CHECK YOUR ANSWERS — If you missed more than two (2) answers, try to correct your mistakes by finding similar problems from this chapter. Then take Review Test #2. Otherwise, ask your instructor for the topic quiz on Decimals II.

CHAPTER 7

Review Test #2

1. 8.09 × .536 = __?__

2. 16.9 ÷ 13 = __?__

3. The product of .474 and .56 is __?__

4. The quotient of 3.98 and 20 is __?__

5. 7.6 × .583 × 12.1 = __?__

6. 78 ÷ .002 = __?__

Use shortcut method in 7–9.

7. 187.9 ÷ 100 = __?__

8. .6321 × 10,000 = __?__

9. .51 × 100 = __?__

10. Volume of a rectangular box is equal to the length of the box times the width of the box times the height of the box ($V = lwh$). If the length is 8.9 cm, the width is 4.3 cm, and the height is 3.7 cm, what is the volume of the box? (The units for volume in this problem are cubic centimeters or cm^3.)

CHECK YOUR ✓ ANSWERS

If you missed more than two (2) answers, correct your mistakes, study the chapter again, and take Review Test #3. Otherwise, ask your instructor for the topic quiz on Decimals II.

CHAPTER 7

Review test #3

1. 28.6 × 13.95 = __?__

2. 3.19 ÷ .319 = __?__

3. The product of 18.74 and .3 is __?__

4. The quotient of .9 and 1.5 is __?__

5. 3.632 × 12 × 7.65 = __?__

6. .2301 ÷ 3 = __?__

Use shortcut method in 7–9.

7. .379 ÷ 100 = __?__

8. 186 × 100 = __?__

9. 7.9637 × 1,000 = __?__

10. Area of a rectangle is equal to length times width ($A = lw$). If the length is 4.98 inches, and the width is .753 inches, what is the area. (Units for area in this problem will be square inches or inches^2.)

CHECK YOUR ANSWERS If you missed more than two (2) answers, *see your instructor* as soon as possible for further assistance. Otherwise, you should be ready to ask your instructor for the topic quiz on Decimals II.

CHAPTER 8

PERCENTS

INTRODUCTION

In this chapter, you will learn the definition of "percent"; how to change percents into decimals and fractions; and how to change fractions and decimals into percents. When necessary use the rounding rule "to the nearest thousandth."

SECTION 8-1 "GETTING OUT OF PERCENT"

What is "percent"? "Percent" means "divided by 100" and is symbolized by this sign: %. Percent is a way of expressing a ratio where the denominator is *always* 100.

EXAMPLE

10% means 10 divided by 100 or $\dfrac{10}{100}$ (What is $\dfrac{10}{100}$ as a decimal? Remember the shortcut for dividing by 100?)

25% means 25 divided by 100 or $\dfrac{25}{100}$ (What is $\dfrac{25}{100}$ as a decimal?)

37.5% means 37.5 divided by 100 or $\dfrac{37.5}{100}$ (What is $\dfrac{37.5}{100}$ as a decimal?)

100% means 100 divided by 100 or $\dfrac{100}{100}$ or 1.

125% means 125 divided by 100 units or $\dfrac{125}{100}$ (What is $\dfrac{125}{100}$ as a decimal?)

Quite often we need to convert percent numbers back into their original decimal or fraction form. You usually do not use percent numbers to multiply or divide other numbers. Instead, the decimal or fraction equivalent of the percent number is used. Therefore, we must now learn how to change percent numbers back into usable decimals and fractions.

Recall that to change fractions into decimals and decimals into fractions was covered in Chapter 6 (see 6-2):

$$.256 = \dfrac{256}{1000} = \underline{\underline{\dfrac{32}{125}}}$$

$$\dfrac{3}{8} = 8\overline{)3.000} = \underline{\underline{.375}}$$
$$\phantom{\dfrac{3}{8} = }\; .375$$
$$\phantom{\dfrac{3}{8} = 8)}\underline{2\ 4}$$
$$\phantom{\dfrac{3}{8} = 8)}\;\;60$$
$$\phantom{\dfrac{3}{8} = 8)}\;\;\underline{56}$$
$$\phantom{\dfrac{3}{8} = 8)}\;\;40$$
$$\phantom{\dfrac{3}{8} = 8)}\;\;\underline{40}$$

Try these for practice to make sure you remember the procedures:

Find the missing number.

	Decimal		Fraction			Decimal		Fraction
1.	.05	=	5/100	8.	.5	=	$\frac{1}{2}$
2.	.25	=	25/100	9.	.75	=	$\frac{3}{4}$
3.	.375	=	375/1000	10.	.375	=	$\frac{3}{8}$
4.	.8	=	8/10	11.	.33	=	$\frac{1}{3}$
5.	.75	=	75/100	12.	.55	=	$\frac{5}{9}$
6.	2.5	=	2 5/10	13.	.8	=	$\frac{4}{5}$
7.	6	=	6/1	14.	1 1/4 = 1.25	=	$\frac{5}{4}$

Check your answers *now*. Do not continue on until *after* you check your answers.

ANSWERS

1. $\frac{5}{100} = \frac{1}{20}$
2. $\frac{25}{100} = \frac{1}{4}$
3. $\frac{375}{1000} = \frac{3}{8}$
4. $\frac{8}{10} = \frac{4}{5}$
5. $\frac{75}{100} = \frac{3}{4}$
6. $2\frac{5}{10} = 2\frac{1}{2}$
7. $\frac{6}{1}$
8. 0.5
9. 0.75
10. 0.375
11. 0.333
12. 0.556
13. 0.8
14. 1.25

Now let's introduce a new column: Find the missing numbers.

	Percent		Decimal		Fraction
1.	15%	=	_____	=	_____

We are now being asked to change 15% back into usable form—that is back into its decimal and fraction equivalents. Remember:

15% means 15 divided by 100 units.

Using the shortcut method for dividing by 100 (7-4), $15 \div 100$ becomes .15

$$15\% \div 100 = .15$$

Problem # 1 from above should now look like this:

	Percent		Decimal		Fraction
1.	15%	=	.15	=	_____

Now all we need is the original fraction that represents 15%. There are several ways to come up with the missing equivalent fraction. One way is to read the decimal, .15, as "fifteen hundredths" (see 6-2). Since $\frac{15}{100}$ is reduceable, we get $\frac{3}{20}$ as the fraction. Another way is to go back to the percent and take another look at what percent means.

$$15\% \text{ means } 15 \text{ divided by } 100 \text{ or } \frac{15}{100}$$

Reduce this fraction and we get the same equivalent fraction, $\frac{3}{20}$. So:

$$\frac{15}{100} = \frac{3}{20}$$

We now have our missing fraction.

		Percent		Decimal		Fraction	
∴	1.	15%	=	.15	=	$\frac{15}{100}$	$= \frac{3}{20}$

Let's try another problem.

	Percent		Decimal		Fraction
2.	36.5%	=	_____	=	_____

Again we are given the percent number and are being asked to find its decimal and fractional equivalents.
What does 36.5% mean?
36.5% means 36.5 divided by 100 or $\frac{36.5}{100}$. At this point it is probably easiest to do the shortcut division by 100 to get .365 ($36.5 \div 100$) which is our decimal equivalent.

2. 36.5% = __.365__ = _____

Now all that's left is to find the equivalent fraction. Since we have the equivalent decimal, just read it to make the fraction. ∴ .365 = $\frac{365}{1000}$ and reduce.

∴ 2. 36.5% = __.365__ = $\frac{365}{1000} = \frac{73}{200}$

IMPORTANT—*Never* "read" the percent number to make the equivalent fraction. Percent numbers should always "look" 100 times larger than either its original decimal or fraction because of the definition of percent. Therefore, reading the percent number merely gives you the *fraction percent*, not the *original* fraction.

You must either divide the percent number by 100 (the meaning of percent) or read the *original* decimal number (the number that appears in the decimal column) to get the equivalent fraction.

Percent   **Decimal**   **Fraction**

3. .01% = _____ = _____

Again we are being asked to convert a percent number (.01%) back into its equivalent decimal and fraction form. Since .01% means $\frac{.01}{100}$, and since we learned that division of a decimal by 100 is simply performed by moving the decimal point two places to the left, we can immediately get our equivalent decimal. (Go ahead and divide .01 by 100 if you wish.)

3. .01% = __.0001__ = _____

Now all that's left is to get the equivalent fraction. Since .0001 is read "one ten thousandth," our fraction is $\frac{1}{10,000}$.

∴ 3. .01% = __.0001__ = $\frac{1}{10,000}$

Percent   **Decimal**   **Fraction**

4. 235% = _____ = _____

235% means $\frac{235}{100}$ which equals 2.35 when you perform the shortcut method for division by 100 (or the long method).

4. 235% = __2.35__ = _____

2.35 is read "two and thirty-five hundredths" which is $2\frac{35}{100}$.

∴ 4. 235% = __2.35__ = $2\frac{35}{100} = 2\frac{7}{20}$

214 CHAPTER 8

> **IN GENERAL**—When given a percent number, state what the percent means (i.e., $\frac{\% \text{ number}}{100}$). This takes the number out of percent and gives you the original equivalent fraction or the original equivalent decimal. Then either read the equivalent decimal to get the equivalent fraction or change the equivalent fraction into the equivalent decimal.

           Percent   **Decimal**   **Fraction**

5. $\frac{1}{6}\%$ = _____ = _____

This may appear strange at first, but go ahead and state what this percent number means:

$\frac{1}{6}\%$ means $\frac{1}{6}$ divided by 100. This time we cannot use our shortcut for division by 100.

(We have no decimal point in $\frac{1}{6}$.) Therefore, we must perform a *fraction division*.

$$\frac{1}{6} \div 100 = \frac{1}{6} \div \frac{100}{1} = \frac{1}{6} \times \frac{1}{100} = \underline{\underline{\frac{1}{600}}}$$

The answer we get after performing the fraction division (don't forget the rules for dividing fractions—see 5-3 and 5-4) is our equivalent fraction.

5. $\frac{1}{6}\%$ = _____ = $\underline{\frac{1}{600}}$

Now all that's left is to change our equivalent fraction, $\frac{1}{600}$, into decimal form.

$\frac{1}{600}$ = $600 \overline{)1.0000}^{.001\overline{6}}$

                600
              4000
              3600
               400 ←

NOTICE: We change the *equivalent fraction* ($\frac{1}{600}$) into decimal form, *not* the percent fraction.

IN GENERAL: Change the given percent fraction into its equivalent fraction first. Then change the *equivalent fraction* into its equivalent decimal.

Remainder keeps repeating ∴ the 6 in our quotient will repeat. Rounded to the nearest thousandth gives us .002.

5. $\frac{1}{6}\%$ = $\underline{.002}$ = $\underline{\frac{1}{600}}$

           Percent   **Decimal**   **Fraction**

6. $\frac{5}{13}\%$ = _____ = _____

Again, state what $\frac{5}{13}\%$ means:

$$\frac{5}{13}\% \text{ means } \frac{5}{13} \div 100$$

Now proceed with the fraction division.

$$\frac{5}{13} \div \frac{100}{1} = \frac{\overset{1}{\cancel{5}}}{13} \times \frac{1}{\underset{20}{\cancel{100}}} = \frac{1}{260}$$

6. $\quad \frac{5}{13}\% \quad = \quad \underline{} \quad = \quad \underline{\frac{1}{260}}$

Change $\frac{1}{260}$ into its equivalent decimal:

$$\frac{1}{260} = 260 \overline{)\begin{array}{l}.00384\ldots \\ 1.00000 \\ \underline{780} \\ 2200 \\ \underline{2080} \\ 1200 \\ \underline{1040} \\ 160 \end{array}} \leftarrow \text{No sign of terminating or repeating.}$$

$\therefore$ 6. $\quad \frac{5}{13}\% \quad = \quad \underline{.004} \quad = \quad \underline{\frac{1}{260}}$

EXERCISE 8-1

Find the missing numbers.

	Percent		Decimal		Fraction
*1.	45%	=	_____	=	_____
2.	9%	=	_____	=	_____
3.	87%	=	_____	=	_____
*4.	156%	=	_____	=	_____
5.	483%	=	_____	=	_____
6.	48.9%	=	_____	=	_____
7.	3.2%	=	_____	=	_____
8.	19.1%	=	_____	=	_____
9.	25%	=	_____	=	_____
*10.	$\frac{1}{9}$%	=	_____	=	_____
11.	$\frac{3}{8}$%	=	_____	=	_____
*12.	$\frac{5}{4}$%	=	_____	=	_____
13.	$\frac{6}{5}$%	=	_____	=	_____
*14.	$12\frac{1}{3}$%	=	_____	=	_____
15.	$83\frac{1}{9}$%	=	_____	=	_____

TRICK: Use the fact that a precent sign, %, looks like a division symbol, to remind you to *always divide percent numbers by 100.*

SECTION 8-2 CHANGING FRACTIONS AND DECIMALS INTO PERCENTS

So far we've looked at changing percent numbers into equivalent fractions and decimals. Now let's see how to change fractions and decimals into percent numbers.

	Percent	Decimal	Fraction	
1.	_____ =	_____ =	$\frac{1}{6}$	← Notice our starting point!

First, we already know how to change fractions into decimals; so getting an answer for the decimals column should be easy:

$$\frac{1}{6} = 6\overline{)1.000}$$

$$\begin{array}{r} .166 \\ \underline{6} \\ 40 \\ \underline{36} \\ 40 \\ \underline{36} \\ 4 \end{array}$$ ← Repeating remainder ∴ repeating digit(s) in answer.

1. _____ = .167 = $\frac{1}{6}$

Now all that needs to be done is to find the equivalent percent number.

> **IMPORTANT**—To change a number into a percent number, multiply the decimal equivalent by 100%.

∴ .167 as a percent is:

$$.167 \times 100\% = \underline{\underline{16.7\%}}$$

You can either remember the shortcut for multiplying by 100 (see 7-2)—move the decimal *two* places to the *right*—or you can actually do the multiplication:

$$\begin{array}{r} .167 \\ \times\ 100\% \\ \hline 16.700\% \end{array} \begin{array}{l} \leftarrow\ \text{3 decimal places} \\ \leftarrow\ +\ \text{0 decimal places} \\ \text{3 total decimal places} \end{array}$$

∴ 1. 16.7 = .167 = $\frac{1}{6}$

	Percent	Decimal	Fraction
2.	_____ =	_____ =	$\dfrac{3}{5}$

Again, the first thing to do is to change the fraction into its equivalent decimal:

$$\frac{3}{5} = 5\overline{)3.0} \quad \begin{array}{r} .6 \\ \underline{30} \end{array}$$

	Percent	Decimal	Fraction
2.	_____ =	.6 =	$\dfrac{3}{5}$

Now, change the decimal (.6) into a percent by multiplying by 100%:

Shortcut or **Long Multiplication**

.6 as a percent = .6 × 100% = <u>60%</u>

$$\begin{array}{r} .6 \\ \times\ 100\% \\ \hline 60.0\% \end{array}$$

∴ 2. <u>60%</u> = <u>.6</u> = $\dfrac{3}{5}$

	Percent	Decimal	Fraction
3.	_____ =	_____ =	$\dfrac{4}{3}$

Change the fraction $\left(\dfrac{4}{3}\right)$ into an equivalent decimal:

$$\frac{4}{3} = 3\overline{)4.000} \quad \begin{array}{r} 1.33\overline{3} \\ \underline{3} \\ 10 \\ \underline{9} \\ 10 \\ \underline{9} \\ 10 \\ \underline{9} \\ 1 \end{array}$$ ← Repeating remainder

3. _____ = <u>1.333</u> = $\dfrac{4}{3}$

Now change the decimal (1.33$\overline{3}$) into a percent by multiplying by 100%:

Shortcut   or   **Long Multiplication**

1.333 as a percent = 1.333 × 100% = 133.3%

$$\begin{array}{r} 1.333 \\ \times \quad 100\% \\ \hline 133.300\% \end{array}$$

∴ 3. $\underline{133.3\%}$ = $\underline{1.333}$ = $\dfrac{4}{3}$

	Percent	Decimal	Fraction
4.	_____	_____	$2\dfrac{5}{16}$

Change $2\dfrac{5}{16}$ into a decimal:

$$2\dfrac{5}{16} = \dfrac{37}{16} = 16\overline{)37.0000}$$

$$\begin{array}{r} 2.3125 \\ \underline{32} \\ 50 \\ \underline{48} \\ 20 \\ \underline{16} \\ 40 \\ \underline{32} \\ 80 \\ \underline{80} \end{array}$$

4. _____ = $\underline{2.3125}$ = $2\dfrac{5}{16}$

Now change 2.3125 into a percent by multiplying by 100%:

Shortcut   or   **Long Multiplication**

2.3125 as a percent = 2.3125 × 100% = $\underline{231.25\%}$

$$\begin{array}{r} 2.3125 \\ \times \quad 100\% \\ \hline 231.2500\% \end{array}$$

∴ 4. $\underline{231.25\%}$ = $\underline{2.3125}$ = $2\dfrac{5}{16}$

	Percent	Decimal	Fraction
5.	_____	= .13 =	_____

This time we are given the decimal and are asked to find the equivalent percent and equivalent fraction.

You can do either first!

You can first read the given decimal to make the fraction or you can first multiply the given decimal by 100% to make the percent number.

$$.13 \text{ is read "thirteen hundredths"} = \frac{13}{100}$$

5. _____ = .13 = $\frac{13}{100}$

$$.13 \text{ as a percent} = .13 \times 100\% = \underline{13\%}$$

∴ 5. __13%__ = .13 = $\frac{13}{100}$

AS A RULE: Do *not* use the fraction equivalent to make the percent number. Instead, always use the decimal equivalent. The multiplication by 100% is easier using the decimal equivalent.

	Percent	Decimal	Fraction
6.	_____	= .005 =	_____

$$.005 \text{ is read "five thousandths"} = \frac{5}{1,000} = \frac{1}{200}$$

6. _____ = .005 = $\frac{5}{1,000} = \frac{1}{200}$

$$.005 \text{ as a percent} = .005 \times 100\% = \underline{.5\%}$$

∴ 6. __.5%__ = .005 = $\frac{5}{1,000} = \frac{1}{200}$

	Percent	Decimal	Fraction
7.	_____ =	4 =	_____

Notice, this time the "decimal equivalent" does not have a decimal point. That's because it's a whole number. However, we could put a decimal point at the right end of the number (4 = 4.) if we wanted to.

To make a whole number look like a fraction, place the whole number over 1.

	Percent	Decimal	Fraction
7.	_____ =	4 =	$\frac{4}{1}$

Now multiply 4 by 100% to make the percent number:

$$4 \text{ as a percent} = 4 \times 100\% = 400\%$$

∴ 7. __400%__ = 4 = $\frac{4}{1}$

	Percent	Decimal	Fraction
8.	_____ =	6.7 =	_____

6.7 is read "six and seven tenths" = $6\frac{7}{10}$

	Percent	Decimal	Fraction
8.	_____ =	6.7 =	$6\frac{7}{10}$

6.7 as a percent = 6.7 × 100% = __670%__   (Use either the shortcut *or* the long multiplication by 100% to get the answer.)

∴ 8. __670%__ = 6.7 = $6\frac{7}{10}$

RULES TO REMEMBER

1. To get out of percent, always *divide* the percent number by 100%, using either the shortcut for division by 100 or long division.

2. To make a percent number, *multiply* the decimal equivalent by 100%, using either the shortcut for multiplying by 100 or long multiplication.

Keep in mind that when you *divide* a number by 100, the answer should look *smaller* than the original number. Also, when you *multiply* a number by 100, the answer should look *larger* than the original number.

EXERCISE 8-2

Find the missing numbers.

	Percent		Decimal		Fraction
*1.	_____	=	_____	=	$\frac{2}{3}$
2.	_____	=	_____	=	$\frac{1}{5}$
3.	_____	=	_____	=	$\frac{5}{6}$
4.	_____	=	_____	=	$\frac{7}{8}$
5.	_____	=	_____	=	$\frac{13}{16}$
*6.	_____	=	_____	=	$\frac{4}{3}$
*7.	_____	=	_____	=	$3\frac{1}{9}$
8.	_____	=	_____	=	$2\frac{5}{8}$
9.	_____	=	_____	=	$\frac{10}{7}$
10.	_____	=	_____	=	$\frac{11}{8}$
11.	_____	=	.25	=	_____
*12.	_____	=	.875	=	_____
13.	_____	=	.5	=	_____
14.	_____	=	.375	=	_____
15.	_____	=	.444	=	_____
16.	_____	=	.008	=	_____
*17.	_____	=	3.75	=	_____
*18.	_____	=	12	=	_____
19.	_____	=	2.4	=	_____
20.	_____	=	6.2	=	_____

CHECK YOUR ANSWERS

You should now be ready for the Chapter Review Test.

CHAPTER 8

REVIEW TEST #1

	Percent		Decimal		Fraction	
1.	34%	=	_____	=	_____	(see 8-1)
2.	_____	=	.58	=	_____	(see 8-2)
3.	_____	=	_____	=	$\frac{3}{5}$	(see 8-2)
4.	16.5%	=	_____	=	_____	(see 8-1)
5.	_____	=	.9	=	_____	(see 8-2)
6.	_____	=	_____	=	$\frac{7}{8}$	(see 8-2)
7.	.13%	=	_____	=	_____	(see 8-1)
8.	_____	=	6.5	=	_____	(see 8-2)
9.	_____	=	_____	=	$\frac{9}{7}$	(see 8-2)
10.	$\frac{5}{8}$%	=	_____	=	_____	(see 8-1)

CHECK YOUR ✓ ANSWERS — If you missed more than four (4) answers (two entire problems), study the chapter again, correct your mistakes, and take Review Test #2. Otherwise, ask your instructor for the quiz on Percents.

CHAPTER 8

REVIEW TEST #2

	Percent		Decimal		Fraction
1.	_____	=	.8125	=	_____
2.	_____	=	_____	=	$\frac{3}{2}$
3.	$2\frac{6}{11}\%$	=	_____	=	_____
4.	_____	=	15	=	_____
5.	_____	=	_____	=	$3\frac{1}{3}$
6.	128%	=	_____	=	_____
7.	_____	=	.0095	=	_____
8.	_____	=	_____	=	$\frac{13}{16}$
9.	.6%	=	_____	=	_____
10.	_____	=	12.4	=	_____

CHECK YOUR ✓ ANSWERS

If you missed more than four (4) answers (two entire problems), study the chapter again, correct your mistakes, and take Review Test #3. Otherwise, ask your instructor for the topic quiz on Percents.

CHAPTER 8

Review Test #3

	Percent		Decimal		Fraction
1.	_____	=	_____	=	$2\frac{2}{3}$
2.	$\frac{7}{9}\%$	=	_____	=	_____
3.	_____	=	.25	=	_____
4.	_____	=	_____	=	$\frac{5}{1}$
5.	20%	=	_____	=	_____
6.	_____	=	8	=	_____
7.	_____	=	_____	=	$\frac{8}{4}$
8.	.43%	=	_____	=	_____
9.	_____	=	.0505	=	_____
10.	_____	=	_____	=	$\frac{5}{14}$

CHECK YOUR ✓ ANSWERS If you missed more than four (4) answers, *ask your instructor* for help as soon as possible. Otherwise, you should be ready to take the topic quiz on Percents.

ARITHMETIC REVIEW

Chapter 1

Whole Numbers {0, 1, 2, 3, 4, . . .}

A. Know how to add, subtract, multiply, and divide.

Chapter 2

Symbols and Definitions

A. Know the words that represent the four basic operations of $+, -, \times, \div$

  Example: Quotient means divide ($\div$).

B. Know how to turn numbers into exponential notation

  Example: $6 \cdot 6 \cdot 7 \cdot 8 \cdot 8 \cdot 8 = 6^2 \cdot 7 \cdot 8^3$

  and how to simplify numbers in exponential notation.

  Example: $3^3 \cdot 4^2 = 3 \cdot 3 \cdot 3 \cdot 4 \cdot 4 = \underline{432}$

C. Know how to identify a rational or irrational number.

  Example: $\sqrt{3}$ = irrational number since it *cannot* be expressed as the ratio of two whole numbers.

D. Know the *order of operations:*

  1. Remove Grouping Symbols.
  2. Perform exponents.
  3. Multiply and divide *in order* from left to right.
  4. Add and subtract *in order* from left to right.

  Example: $3^3 + 8(4 - 1) - 2^3 \div 2(3 + 7)$

  $3^3 + 8 \ (3) \ - 2^3 \div \ 2(10)$

  $27 + 8(3) \quad - 8 \div 2(10)$ **REMEMBER:** 8(3) means "multiply 8 times 3."

  $27 + 24 \quad\quad - 4(10)$

  $27 + 24 \quad\quad - 40$

  $51 \quad\quad\quad - 40$

  11

Chapter 3

Prime Numbers: *Prime*—whole number larger than 1, divisible only by 1 and itself.

A. Know how to "break" a number into its prime factors through use of "factor tree."

Example:

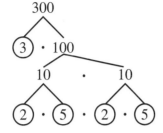

So, 300 = 2 • 2 • 3 • 5 • 5

B. Know how to determine LCM of 2 or more numbers. Find prime factors of each number then take greater or greatest number of appearances of a given prime to help form LCM.

Example: Find LCM for 24, 60, 28

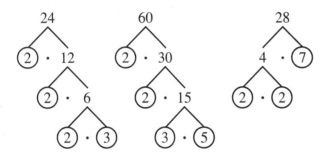

24	2	2	2	3		
60	2	2		3	5	
28	2	2				7

LCM = 2 • 2 • 2 • 3 • 5 • 7 = 840

Chapters 4 and 5

Fractions

A. *Add and Subtract*—To add and subtract fractions, *make sure you have common denominators (use LCM method to find the LCD)*. Subtractions may have borrowings.

Example: $1\frac{5}{8} + 6\frac{2}{9} - 4\frac{13}{15}$

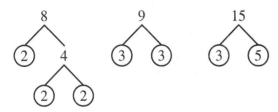

8	2	2	2		
9				3	3
15				3	5

LCD = 2 • 2 • 2 • 3 • 5 • 5 = 360

Don't forget how to "build" equivalent fractions, which is necessary *after* you find the LCD.

$$1\frac{5}{8} = 1\frac{225}{360}$$

$$+ 6\frac{2}{9} = + 6\frac{80}{360}$$

$$7\frac{305}{360} = 6\frac{665}{360}$$

$$- 4\frac{13}{15} = - 4\frac{312}{360} = - 4\frac{312}{360}$$

$$2\frac{353}{360}$$

Example: $1\frac{5}{8} = 1\frac{?}{360}$ since $360 \div 8 = 45$

5×45 is our new numerator.

$$1\frac{5(\times 45)}{8(\times 45)} = 1\frac{225}{360}$$

Answer: $2\frac{353}{360}$

B. *Multiply and Divide*—To multiply and divide fractions, first make sure you have "pure" fractions; that is, *all mixed numbers must become improper fractions.*

Example: $3\frac{4}{7} = \frac{25}{7}$

To divide, invert (flip over) the fraction following the division sign (the divisor) and then multiply.

Example: $\frac{8}{9} \div \frac{5}{7} = \frac{8}{9} \times \frac{7}{5}$ or $\dfrac{\frac{8}{9}}{\frac{5}{7}} = \frac{8}{9} \div \frac{5}{7} = \frac{8}{9} \times \frac{7}{5}$

- Cancellation among or between fractions can only be performed while multiplying.
- To multiply fractions, multiply the numerators and multiply the denominators to form a new numerator and a new denominator, respectively.

Example: $\dfrac{4}{11} \times \dfrac{15}{24} \times \dfrac{22}{35} = \dfrac{\cancel{4}^1}{\cancel{11}_1} \times \dfrac{\cancel{15}^3}{\cancel{24}_6} \times \dfrac{\cancel{22}^2}{\cancel{35}_7} = \dfrac{1}{1} \times \dfrac{1}{1} \times \dfrac{1}{7} = \dfrac{1}{7}$

Example: $\dfrac{12}{8} \times \dfrac{8}{3} \times \dfrac{15}{9} = \dfrac{\cancel{12}^4}{\cancel{8}_1} \times \dfrac{\cancel{8}^1}{\cancel{3}_1} \times \dfrac{\cancel{15}^5}{\cancel{9}_3} = \dfrac{4}{1} \times \dfrac{1}{1} \times \dfrac{5}{3} = \dfrac{20}{3} = 6\dfrac{2}{3}$

Example:* $1\dfrac{5}{6} \times 2\dfrac{4}{7} = \dfrac{11}{6} \times \dfrac{18}{7} = \dfrac{11}{\cancel{6}_1} \times \dfrac{\cancel{18}^3}{7} = \dfrac{33}{7} = 4\dfrac{5}{7}$

Example:* $4\dfrac{2}{3} \div 5\dfrac{8}{9} = \dfrac{14}{3} \div \dfrac{53}{9} = \dfrac{14}{\cancel{3}_1} \times \dfrac{\cancel{9}^3}{53} = \dfrac{42}{53}$

When multiplying and dividing, *all* mixed numbers must be turned into improper fractions.

Chapters 6 and 7

Decimal Numerals (Decimals)

A. Be able to write a decimal as a fraction and vice-versa.

Example: $85.4 = 85\frac{4}{10}$; $6\frac{145}{1000}$; $.8937 = \frac{8937}{10,000}$

Example: $\frac{8}{5} = 5\overline{)8.000} = 1.6$; $\frac{2}{3} = 3\overline{)2.00000} = .667$

$$\begin{array}{r} 1.6 \\ 5\overline{)8.000} \\ \underline{5} \\ 30 \end{array} \qquad \begin{array}{r} .666\ldots \\ 3\overline{)2.00000} \\ \underline{18} \\ 20 \\ \underline{18} \\ 20 \\ \underline{18} \\ 20 \end{array}$$

1. Be able to round off to a given position.

   Example: Round 8.345678 to the nearest thousandth.

   8.345 678   (5 is in the thousandths place and 6 is large enough [5 or larger] to increase the thousandths place by 1.)

   Answer: 8.346

B. Addition and Subtraction of Decimals—Be sure to *line up the decimal points* and add zeros if necessary.

   Example: $45.7 + 1.725 = \begin{array}{r} 45.7 \\ +1.725 \end{array} = \begin{array}{r} 45.700 \\ +1.725 \\ \hline 47.425 \end{array}$

   Example: $11.14 - 9.346 = \begin{array}{r} 11.14 \\ -9.346 \end{array} = \begin{array}{r} 11.140 \\ -9.346 \\ \hline 1.794 \end{array}$

   If a number has "no" decimal, we know it's at the end (far right) of the number.

   Example: 190 = 190.

   Example: $27 - 3.44 = \begin{array}{r} 27. \\ -3.44 \end{array} \quad \begin{array}{r} 27.^{1}0^{1}0 \\ -3.44 \\ \hline 23.56 \end{array}$

C. Multiplication of Decimals—Multiply the numbers as if the decimals were not present (as if multiplying two whole numbers); then *count* the number of decimal positions filled by the two numbers and, starting at the right end of your answer, count that many places to the left and locate your decimal point.